DRIVER CPC

A GUIDE

FOR DRIVERS OF BUSES AND COACHES

Licences D, D+ E, D1 and D1 + E

WRITTEN AND COMPILED BY

HARDINGTON COMMUNICATIONS LTD

Published by Hardington Communications Ltd

Copyright © Hardington Communications Ltd 2014

Written and compiled by Hardington Communications Ltd

© Copyright 2014

All rights reserved

Acknowledgements

We would like to thank the following for their support in producing this publication.

Charlotte Sears and Ainley Wade without whose support this this project would not have been possible.

Ilustrations by ami artwork.

The Society of Operations Engineers & Institute of Road Transport Engineers for their data..

All our friends at South West Coaches, Taylor's Coaches, Lufton Commercials, Bakers Coaches and Somerset School Transport.

We would also like to thank all the relevant government departments for allowing the use of material under Open Government Licence v1.0.

No warranty

Whilst every effort has been made to ensure that the information in this publication is as accurate as possible, the information cannot be guaranteed against errors or omissions. The information provider shall not be liable for any loss, injury or damage of any kind caused by the use of this publication.

CONTENTS

WHAT IS DRIVER CPC?..1
WORKING TIME DIRECTIVE, THE ROAD TRANSPORT DIRECTIVE, THE HORIZONTAL AMENDING DIRECTIVE..4
DRIVERS' HOURS...9
TACHOGRAPHS..17
UNDERSTANDING YOUR VEHICLE AND ITS CHARACTERISTICS..........27
MORE EFFICIENT USE OF FUEL...30
VEHICLE SAFETY AND WALK AROUND CHECKS................................33
WHEELS AND TYRES..39
DIMENSIONS AND WEIGHTS OF VEHICLES......................................43
TRANSMISSION SYSTEMS..47
COMMERCIAL VEHICLE BRAKING SYSTEMS....................................51
UNDERSTANDING THE IMPORTANCE OF HEALTH AND WELL BEING WHEN DRIVING...55
BASIC FIRST AID TECHNIQUES..63
MANUAL HANDLING..71
PASSENGER SAFETY AND COMFORT...76
FORCES AFFECTING YOUR VEHICLE AND PASSENGERS....................81
SEAT BELT REGULATION..86
THE CARRIAGE OF PEOPLE WITH DISABILITIES OR SPECIAL NEEDS...90
HOW TO ASSESS AND DEAL WITH EMERGENCY SITUATIONS...........94
PROPOSED EVACUATION PROCEDURE..98
MANAGING YOUR PASSENGERS AND CONFLICTS..........................103
CRIMINALITY AND ITS EFFECTS...111
VEHICLE SECURITY CHECK LIST AT BORDER CROSSINGS................115
CARRIAGE OF PASSENGERS IN THE UK AND INTERNATIONALLY......116
UNDERSTANDING THE RISKS WHEN WORKING AND DRIVING.........121
REPORTING A ROAD TRAFFIC INCIDENT.......................................129
ENHANCING THE IMAGE OF YOUR COMPANY...............................131
ECONOMICS AND THE TRANSPORT INDUSTRY..............................136
APPENDICES...142
LICENCE CATEGORIES FOR DRIVING PCVS....................................143
SPEED LIMITS...145
COUNTY TICKET AND SCHOOL BUS PASS GUIDE...........................146
ANSWERS TO QUESTIONS..147

WHAT IS DRIVER CPC?

Driver CPC is a **'Certificate in Professional Competence'** that most professional drivers will need to get in order to drive for a living. The qualification is not just for the UK; it is a European law and covers all countries in the EU.

What is its purpose?
The purpose of the Driver CPC is to improve the professional standards of drivers throughout Europe. It will confirm or expand the existing knowledge and understanding of drivers. It is believed that in doing so, it will improve road safety, save lives and reduce injury as well as improve the environment through more fuel efficient driving.

Who needs a CPC?
Most professional drivers will need to get a CPC in order to drive for a living. The Driver CPC covers most licence categories, so drivers of buses, mini-buses and coaches (D1, D1E, D and DE) and large vans, small and large lorries (C1, C1E, C and CE) will need to have their driver CPC qualification.

What if I have more than one category of driving licence?
If you hold a licence to drive both coaches and lorries (DE and CE licences), you only need to complete 5 days or 35 hours of training. You <u>don't</u> have to do separate training for each licence category or type of vehicle.

Are there any exemptions?
There are exemptions. The following is a list of vehicles whose drivers do not have to obtain their CPC qualification:
- Drivers of vehicles that have a maximum authorised speed of 45 Km/h
- Vehicles used under the control of the armed forces, civil defence, the fire service and forces with responsibility for maintaining public order
- Vehicles undergoing a road test for technical development or repair
- Vehicles used in states of emergency or on rescue missions
- Vehicles used in the course of driving lessons
- Vehicles used for personal use and the non-commercial carriage of goods or people

Vehicles used for the carriage of goods or equipment to be used by the driver provided that driving that vehicle is not the driver's principal activity.
N.B. This is not the complete list of exemptions.

When does it start?
It has already started. All new bus drivers (D1, D and DE licences) have been required to hold their CPC Driver Qualification Card from September 2008 and all new lorry drivers (C1, C and CE licences) have had to have their CPC since September 2009. Existing bus drivers have until September 2013 and existing lorry drivers have until September 2014 to get their Driver CPC.

What do I have to do to obtain a CPC?
In order to get your CPC qualification you will need to attend 5 days or 35 hours of what is called periodic training. It is called periodic training as it can be done over a 5 year period. Usually one day a year. All training must be done with an approved Driver CPC trainer.

What if I don't get the Drivers' CPC qualification?
If you don't get your Driver's CPC qualification and are intending to drive a vehicle not listed in one of the exemptions categories. When you reach the deadline for that category of vehicle, you will not be allowed to drive and your driver's licence will be invalid for non-exempted vehicles.

What If I start but don't complete the course?
If you don't complete the you training course by the deadline set (September 2013 for Coach and Bus drivers and September 2014 for Lorry and Large Van drivers) you will not be able to drive those vehicles legally.

What if I am not a UK citizen?
You can complete the Driver CPC course in the UK. Any training completed in another EU country will count towards the UK Driver CPC. You must keep your certificate safe as it is the only proof you will have to show that you have attended the course.

How long is the Driver's CPC qualification valid for?
Your Driver CPC will last 5 years from September 2013 for Buses (September 2018) and for Lorries it will be 5 years from September 2014 (September 2019). If you completed the course today you wouldn't need to obtain the next Driver CPC until 2018 or 2019.

What happens when I complete my training?
As you complete each 7 hours' training you will receive a certificate confirming your training from your training provider. Your trainer will then register your training with DSA and DVLA. So when you have completed all 5 days or 35 hours you will automatically receive your Driver Qualification Card (DQC). You must then carry this card when working.

Where can I get trained?
In most cases your employer will arrange the training for you. However a list of training organisations for your area can be found on the JAUPT website. Go to www.drivercpc-periodictraining.org

Where can I get more information?
Further information can be obtained from customerservices@dsa.gsi.gov.uk or telephone 0300 200 1122.

WORKING TIME DIRECTIVE (WTD)
THE ROAD TRANSPORT DIRECTIVE (RTD)
THE HORIZONTAL AMENDING DIRECTIVE (HAD)

In addition to all the rules relating to drivers' hours coming from the EU, there are other sets of rules shown above. The legislation, which affects most employees not just drivers, was brought in to try and give better protection for workers within the EU. In particular they seem to have targeted the working conditions of night workers.

Originally the directive (WTD) excluded the transport sector. However in 2003 and 2005 the regulations were changed and this brought in new rules which now include drivers of most types of vehicles. In 2009 further changes were brought in which allowed for the inclusion of self employed drivers. There are currently minor differences in each of the directives. However, it is expected that these differences will merge in time.

What is 'working time'?
This is defined as time spent at work undertaking activities or duties. This includes travelling as part of the job, but not the time travelling to and from work, rest or break times or periods of availability (POAs) time on call at home. It does however include waiting time to load or unload or time waiting for ferries.

What are the regulations?
The Working Week
The working week starts at 00.00 Monday to 24.00 Sunday. The driver must not exceed an average of 48 hours, including overtime, over a 17 week period, with a maximum of 60 hours in any one week. That said, this can be extended to an average of 48 hours based over 26 weeks, by agreement with your union or other works committee. Statutory holiday time (20 days per year) is included and given a notional allowance of 8 hours per day or 48 hours per week. The example shown is of a rolling reference starting at the start of week 4 to the end of week 20. A fixed calendar reference starting in April can also be used.

		10 hours overtime for 12 weeks		
		Standard 40 hours working week for 15 weeks		2 weeks holiday
Week	1 2 3	4 5 6 7 8 9 10 11 12 13 14 15	16 17 18	19 20

From the example shown above we can see that the driver has worked a total of 15 weeks plus two weeks holiday starting at week 4 and running to week 20. The number of hours calculated under the working time directive will be:

$$10 \times 12 = 120 \text{ overtime}$$
$$\text{Plus } 40 \times 15 = 600 \text{ working week}$$
$$\text{Plus } 48 \times 2 = 96 \text{ holiday time}$$
$$\text{The total hours} = 816 \text{ hours}$$

The average calculated over the 17 week period is therefore 48 hours and the 60 hour maximum has not been exceeded in any one week.

Night Working
For the purpose of the regulations night work is defined as the period between 12.00 am and 4.00 am for LGV drivers and 12.00 am to 5.00 am for PCV drivers. Night workers are limited to 10 hours work in any 24 hour period under RTD.

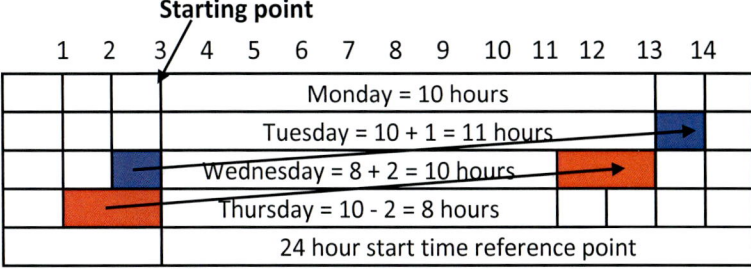

The rules apply to each 24 hour period. So with a reference starting point at 3.00 a.m. Tuesday, the earlier start of 1 hour on Wednesday is added to Tuesday's hours. Likewise Thursday's 2 hours early start is added to Wednesday's hours, leaving only 8 hours recorded for Thursday. Each period of 24 hours is calculated separately so if a driver changes to day work later that week the rules will no longer apply.

Breaks
Under the Road Transport Directive (RTD) drivers are entitled to a break or breaks totalling 30 minutes (if working between 6 and 9 hours) and 45 minutes (if working over 9 hours).

Daily and Weekly Rest Periods
Employers are required to allow staff to take a daily rest period of 11 consecutive hours in every 24 hours and a weekly rest period of not less than 24 hours every 7 days.

Holidays
Full time employees are entitled to 4 weeks (20 days) paid holiday per year. This is in addition to national holidays.

Health Assessments
Employers have a duty to provide free health assessments for night workers.

Keeping of Records
Employers are required to keep records of hours that their employees work. These records must be kept for a period of 2 years from the date they were produced and must be available for inspection by VOSA. It is advisable to keep records, other than tachograph records, to show time worked by drivers and other employees. Employees have a duty to inform employers in writing of any other work they do for another employer.

It needs to be noted that what is shown here is not definitive, simply an overview and there are many anomalies among the sets of legislation, including those for younger workers, such as apprentices.

Questions on Directives WTD, RTD and HAD

Q1. You spend an hour each day driving to and from work.
   A. This time forms part of your working week and must be recorded
   B. This time is part of your working time but does not need to be recorded
   C. This time does not form part of your working week
   D. This time must be recorded on your time sheet

Q2. Under the RTD, employers are required to keep your time sheets for inspection for a period of...
A. 1 year
B. 18 months
C. 2 years
D. Do not need to keep them

Q3. You are employed on regular night work with your company. Your employer should...
A. Ask you from time to time if you are in good health
B. Keep more detailed records of your hours
C. Offer you a free eye test
D. Ensure that you have regular health checks

Q4. Your employer must keep your time sheets for inspection by...
A. HM Revenue and Customs
B. Department of Trade and Industry
C. VOSA
D. Department of Health

Q5. Whilst waiting for a ferry you take a break of 45 minutes. This time should be...
A. Included as part of your working time
B. Not included as working time
C. Included if you are sat in the vehicle
D. Included if the ferry is in port

Q6. Your average working week when working under the WTD should not exceed...
A. 60 hours
B. 50 hour
C. 45 hours
D. 48 hours

Q7. When employed full time you are entitled to paid holidays. Under the working time agreement the number of weeks that you should receive is...
A. Two weeks or 10 days
B. Three weeks or 15 days
C. Four weeks or 20 days
D. Five weeks or 25 days

Q8. Under the Road Transport Directive (RTD), you are entitled to take a break of 45 minutes if your working day is more than...
A. 6 hours
B. 4 hours
C. 8 hours
D. 9 hours

Q9. Under the directives, employees are entitled to take a daily rest period of at least...
A. 8 hours
B. 12 hours
C. 9 hours
D. 11 hours

Q10. Under the directives, employees are entitled to take a weekly rest period of at least...
A. 48 hours
B. 36 hours
C. 24 hours
D. 30 hours

DRIVERS' HOURS

This topic will be covered in six sections.

1. Domestic Drivers' Hours (PCV Drivers).
2. EU Drivers' Hours.
3. Mixed E.U. & Domestic Drivers' Hours.
4. Multi Manning.
5. Sample Test Questions.

1. Domestic rules for PCV Drivers
Domestic rules only apply to drivers and vehicles on journeys within the UK that have been exempted from EU rules. Also within these rules there are specific differences between the rules for truck and bus (LGVs & PCVs).

Regular Service
Drivers operating on a regular service of up to 50 Km are free from EC rules and generally operate under domestic drivers' hours. No written records are required when operating under these rules. However, if the driver also drives a LGV, he/she must keep a record book to record driving duties etc. Under domestic rules, drivers are not allowed to drive for more than 10 hours in any one day.

Working Day
Under domestic hours the working day must be no more than 16 hours. The driver must have 10 hours rest between any two working days. However, on three occasions during a working week the rest periods may be reduced to 8.5 hours. That said, the working time directives place limits of a maximum working week of 60 hours and an average of 48 hours over a period of six months, but only if agreed to by the driver, otherwise it will be averaged over a 17 week period. Also the number of hours that night workers can work is limited to 10.

Work day		1		2		3		4		5		6	
Rest period	Weekly rest ends		Daily rest 10 hours		Daily rest 8.5 hours		Daily rest 8.5 hours		Daily rest 10 hours		Daily rest 8.5 hours		Weekly rest starts

2. EU Rules

New EU rules on drivers' hours came into force in April 2007. Their purpose was to simplify the rules in existence at that time and provide for further updates in legislation.

Under EU rules the number of hours that a driver may work and the rest periods that he or she must take are generally recorded by using a tachograph. The type of work and any rest periods are shown using the symbols below.

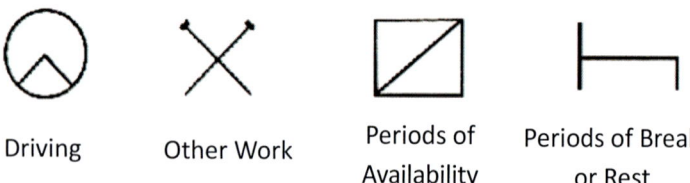

Driving   Other Work   Periods of Availability   Periods of Break or Rest

Daily Driving

Under EU rules, a day is said to be any 24 hour period. Your daily work begins when you either start to work or begin driving after a weekly or daily rest period. The maximum daily hours that you may drive is nine (9). However, this can be increased to ten (10) hours up to twice a week.

Mon	Tues	Wed	Thurs	Fri	Sat	Sun	Total
9 hrs	9 hrs	10 hrs	9 hrs	10 hrs	9 hrs	0 hrs	56 hrs

Weekly Driving

Under EU rules the week begins at 00.00 hours on Monday and ends at 24.00 hours the following Sunday.

00:00 Mon	Tues	Wed	Thurs	Fri	Sat	24:00 Sun

Drivers are restricted to a weekly driving limit of **56 hours**. However, drivers must not exceed a total of 90 hours in any two week period.

	Week 1	Week 2	Week 3
	Two week total = 90 hours		
	40 hours	50 hours	40 hours
		Two week total = 90 hours	

Unforeseen events

An unforeseen event is something that has not been planned for or could not have been predicted beforehand.

The rules state that provided that road safety is not compromised, a departure from the drivers' hours rules is allowed where there is an unforeseen event. However, this must not be a regular occurrence and the reasons for a departure from the rules **must** be recorded when the driver reaches a suitable and safe stopping place. This condition must be applied for reasons of safety only and should not be used to allow a driver to complete his or her journey.

Examples of such events are:
1. Delays caused by extreme weather.
2. Road traffic accidents.
3. Breakdowns.
4. Interruptions of ferry services.

Please note that during a journey that involves the vehicle being driven on and off public roads, the off road time will also count as driving time.

Breaks

You must ensure that you take an uninterrupted break of 45 minutes after driving for four and a half hours (4.5 hours).

Daily rest	4.5 hours driving	45 minutes break	4.5 hours driving	Daily rest
Total of 9 hours driving				

Daily rest	4.5 hours driving	45 minutes break	4.5 hours driving	45 minutes break	1 hour driving	Daily rest
Total of 10 hours driving						

However, a 45 minutes break can be replaced by a break of at least 15 minutes followed by a break of 30 minutes. A 45 minute break or split breaks totalling 45 minutes are required before getting back behind the wheel.

Note. During your break you must not drive or undertake any other work. Breaks must be uninterrupted.

Daily rest	2 hours driving	15 minutes break	2.5 hours driving	30 minutes break
Total of 4.5 hours driving				

2 hours driving	15 minutes break	2.5 hours driving	30 minutes break	1 hour driving
Total of 5.5 hours driving				

Daily Rest Periods
Drivers need to take a daily rest period of at least 11 hours. However, this can be taken in two blocks, one of which must be at least 3 hours and the other 9 hours. The 9 hour period being the last block. Both of which must be uninterrupted. By taking you daily rest in two blocks you increase the daily rest time from 11 to 12 hours.

Monday 9.00 am	4.5 hours driving	3 hours rest	4.5 hours driving	9 hours rest	Tuesday 6.00 am

A reduced daily rest period of 9 hours cannot be taken more than three times between any two weekly rest periods.

Work Day		1		2		3		4		5		6	
Rest Periods	Weekly rest ends		Daily rest 11 hours		Daily rest 9 hours		Daily rest 9 hours		Daily rest 11 hours		Daily rest 9 hours		Weekly rest starts

Weekly Rest Periods
After a maximum of 6 periods of 24 hours, a regular or standard weekly rest period of rest of at least 45 hours should be taken. You can take two regular weekly rest periods in any two consecutive weeks or you can take one regular period of 45 hours and a reduced period of a minimum of 24 hours. However, if for example you take a reduced period of weekly rest of 30 hours, you must add the lost time to another daily or weekly rest period. (see diagram over).

Saturday 10:00	Monday 7:00	↓	Saturday 12:00	Sunday 18:00	↓	Friday 18:00	Monday 06:00
Regular weekly rest of 45 hours			Reduced weekly rest of 30 hours			Extended weekly rest of 60 hours	
	Working week			Working week			

Saturday 11:00	Monday 8:00	↓	Saturday 12:00	Sunday 12:00	↓	Friday 18:00	Monday 12:00
Regular weekly rest of 45 hours			Reduced weekly rest of 24 hours			Extended weekly rest of 66 hours	
	Working week			Working week			

45 hours – 30 hours = 15 hours so you will need to add 15 hours to your next daily or weekly rest period before the end of the third week following the week in question.

We can see in this last example that the reduced weekly rest period of 24 hours is compensated for with an extended weekly rest period. Therefore the minimum rest period you will need to take to catch up is 45 hours + 21 hours = 66 hours. The rest taken as compensation for the reduced weekly rest period must be taken in one block.

Any periods of compensatory rest can be taken away from base in your vehicle, provided that it is stationary and has sleeping facilities.

Weekly rest of 24 hours	⟶			Sunday 24:00
Week 1	Week 2	Week 3	Week 4	

Compensation for the shortened weekly rest period (45 - 24 = 21 hours) must be taken by the end of the third (3) week following the week of the reduced break.

3. Mixed EU and Domestic Rules
Drivers may find themselves working under both sets of regulations. In such situations drivers must apply the appropriate set of rules for the duties that are being undertaken. So under EU rules you will need to use a tachograph and record on the back of the chart or digital print out your domestic hours. Alternatively, you could use the tachograph for both Domestic and EU journeys.

4. Two or more Drivers (Multi Manning)

When there are two drivers in the vehicle, each driver must ensure that their cards or discs are entered into the correct slots in the tachograph. Slot 1. Is for the current driver and slot 2 is for the second or standby driver

The system then records the time that the second driver is in the passenger seat. When the drivers change over, the cards need to be swapped. By having two drivers, the drivers can spread their duty time over a period of 21 hours. Each can have a break whilst the other is driving but not a daily rest period. (Remember the vehicle must be stopped during the daily rest period). Each driver must take a daily rest of not less than 9 consecutive hours. This needs to be done within a 30 hour period.

Questions on Drivers' Hours

Q1. What is the maximum number of hours that a driver may drive in one week under EU rules?
   A. 48 hours
   B. 45 hours
   C. 90 hours
   D. 56 hours

Q2. Under EU rules, how long can a driver drive before he/she must take a break?
   A. 2.5 hours
   B. 4 hours
   C. 6 hours
   D. 4.5 hours

Q3. What is the maximum number of hours that a driver may drive in any two week period under EU rules?
   A. 80 hours
   B. 90 hours
   C. 95 hours
   D. 75 hours

Q4. What is the minimum weekly rest period that a driver must take if at his base under EU rules?
   A. 36 hours
   B. 30 hours
   C. 24 hours
   D. 40 hours

Q5. Under EU rules, after driving for two hours, the minimum time period that a driver may take a break is?
A. 45 minutes
B. 30 minutes
C. 15 minutes
D. 10 minutes

Q6. Under EU rules what is the maximum number of hours that a driver may drive in a daily/24 hour period?
A. 9 hours
B. 10 hours
C. 8 hours
D. 12 hours

Q7. Under EU rules, the normal number of hours that a driver may drive in a daily/24 hour period can be increased twice a week by how much?
A. 2 hours
B. 1 hour
C. 10 hours
D. 9 hours

Q8. Under EU rules if a driver drives for 48 hours during a week 'A', what is the maximum number of hours that he can drive for during week 'B'?
A. 45 hours
B. 40 hours
C. 42 hours
D. 50 hours

Q9. When driving for more than 4 hours in any 24 hour period under domestic hours rules, what is the maximum daily duty time you are allowed to work including your driving time for PCV.
A. 24 hours
B. 16 hours
C. 12 hours
D. 11 hours

Q10. Under EU rules what is the minimum daily rest period that can be taken?
A. 11 hours
B. 9 hours
C. 3 hours
D. 24 hours

Q11. Can a daily rest period be split into two parts under EU rules?
   A. Yes
   B. No
   C. Only at weekends
   D. Only on Fridays

Q12. Under EU rules, how long should my regular weekly rest period be?
   A. 40 hours
   B. 36 hours
   C. 24 hours
   D. 45 hours

Q13. How do I take account of the time difference if I am driving in the UK and overseas?
   A. Just stick to UTC time
   B. Adjust your clock when you get to France
   C. Work only to British Summer Time
   D. Work to the time of the country you are travelling through

Q14. When do domestic drivers' hours apply?
   E. Only when working in the UK
   F. When working off road
   G. When working less than 50 Km from home
   H. When working less than 100 Km from home

Q15. What do I have to do if I am working under a mix of domestic and EU rules?
   A. Keep records of both types of work
   B. Only keep records of EU rules driving
   C. No need to keep any records
   D. Keep domestic records only

TACHOGRAPHS

When driving commercial vehicles for hire and reward in the European Union, drivers' working hours are controlled in the interest of road safety. Controls are also there to ensure fair competition across the whole of the European Union States.

In order to ensure that all commercial drivers are complying with the rules, tachographs are used to measure the types of work and rest periods that a driver takes whilst undertaking his duties as a driver.

The tachograph has the ability to record **Time, Distance** and **Speed** of the vehicle.

As tachographs are a legal requirement, operators and drivers have a duty to ensure that they operate properly. They must be checked and calibrated at regular intervals by trained technicians at approved calibration centres or by the manufacturers.

There are two types of tachograph in operation today. Analogue and Digital. There are also a number of different manufacturers of each of these types:

- **Analogue** tachographs need to be checked every two years and recalibrated every six years.
- **Digital** tachographs need to be recalibrated every two years.

Should a driver find a fault with the tachograph, it must be repaired or replaced by an approved centre as soon as possible. If the vehicle cannot return to base within seven days of the fault being discovered, then it must be repaired or replaced during the journey. Drivers need to be aware that whilst the tachograph is broken, they must keep separate written records of their breaks, rest and driving times.

Analogue Tachographs
This type uses a cardboard card or disc, on which a number of traces are printed out. Drivers must carry enough charts for the whole journey; one for each 24 hour period and also some spares. Your employer is responsible for ensuring you have enough charts.

Drivers have a responsibility to ensure that the correct information is recorded on the chart. This includes:

1. Your surname and first name.
2. The date and place your journey begins and will end.
3. The registration number of the vehicle.
4. The odometer (mileage) at the start and end of the journey or journeys or change of vehicle.
5. The time of the change of vehicle. The time needs to be the official time of the country of registration of the vehicle.

If you change vehicles and the new vehicle has a different type of tachograph, you must take the first chart or card with you in the second vehicle. It is the driver's responsibility to ensure that all the information is recorded correctly.

Also, the operator has responsibility to ensure that drivers are following the rules. Drivers and operators can receive large fines for failing to record or for misuse of the charts.

Inspection of Charts
Drivers must make their charts available for inspection by the authorities and must carry print outs or cards for the current day and the previous 28 days. If you also have a digital smart card, you must carry the card at all times when you are driving.

If the enforcement officer retains your cards, he/she must give you a receipt or endorse the replacement card with:

- His or her name and telephone number.
- The number and the dates of any retained cards.

Analogue Tachograph Disc

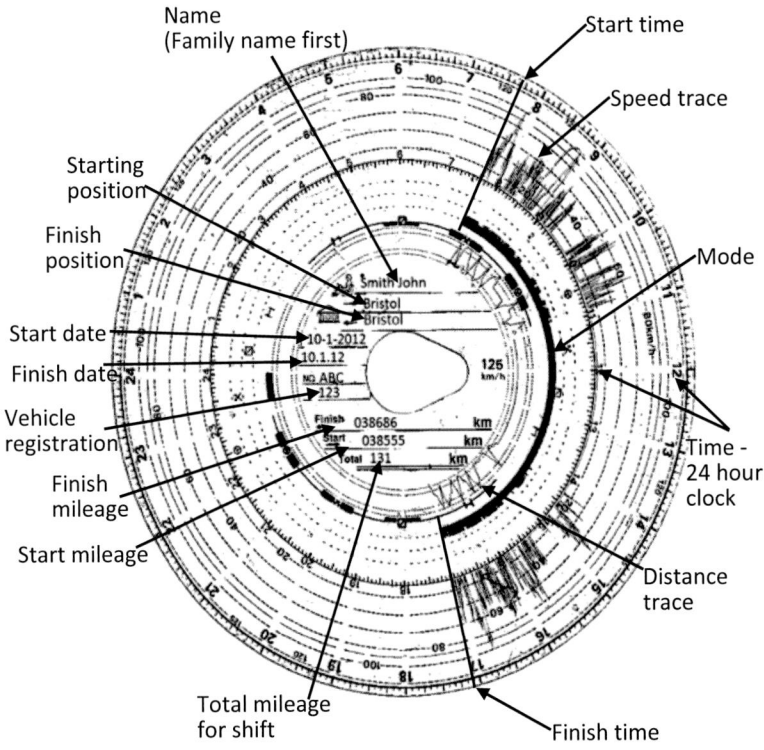

Analogue Tachograph Disc (Manual Entry)

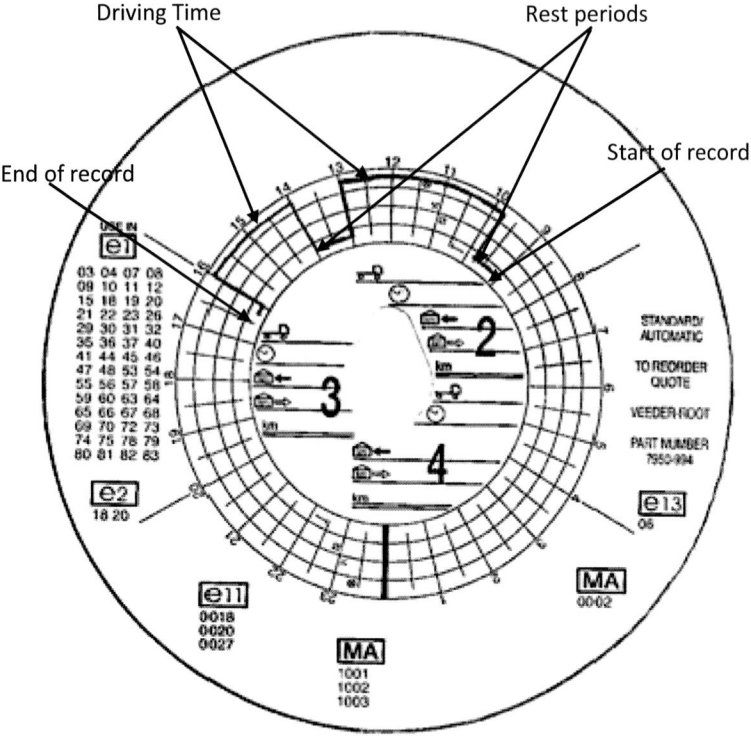

Manual Entry on Analogue Disc
Should the analogue tachograph fail for what ever reason, drivers can record some of the details manually on the rear of the disc.

Here we can see how to produce a manual entry on the rear of the analogue tachograph disc. It can be seen that on this disc the record flows anti-clockwise. The driver has recorded his/her driving time and breaks. The record shows that the driver takes a break between 9.00 am and 9.45 am before starting to drive at 9.45 am until 1.00 pm. At this point the driver takes a second break before he resumes driving until 4.00 pm.

Digital Tachographs

Digital tachographs became mandatory on all new commercial vehicles from May 2006 and have to be calibrated every two years. All tachographs need to be recalibrated if operators change the size of the tyres.

Once drivers place their smart cards into the tachograph, the instrument will record all the driver's details, the duties undertaken, the date and time and the operation of the vehicle. However, if the driver fails to place his/her smart card into the digital tachograph, the tachograph will record that no card has been installed but continue to monitor the operation of the vehicle. Tachograph units can record up to 365 days of data and operators are required to download this data at intervals of at least every three months.

Drivers and operators must be able to make prints of their driving duties if requested. Always ensure that you have sufficient print roll in the vehicle. As with the analogue tachograph, if a driver breaches the drivers' hours rules due to unforeseen circumstances, he/she must print out their activities for that day and make a manual entry on the back of the printout, giving a reason why there was a breach of the rules.

The digital tachograph also has a memory of its own and can record all the details that have been recorded on the driver's smart card. Should the digital tachograph become faulty or the driver lose his/her smart card, he/she will need to make manual entries for each day on the print roll of their duties, dates and times. They also need to identify themselves with name, licence number and then sign the manual entry.

Drivers' Smart Cards

Drivers who have never held a smart card cannot drive vehicles with digital tachographs. If a driver has lost or misplaced his or her card they can continue to drive vehicles with digital tachographs for up to 15 days, provided they make manual entries. However, if they have not obtained a replacement within 15 days, they must stop driving vehicles with digital tachographs.

The card is designed to prevent drivers' hours' offences and is personalised to the driver. So it cannot be used by another driver. The card will be valid for 5 years and include personal information.(See list).

Digital Tachograph Printout

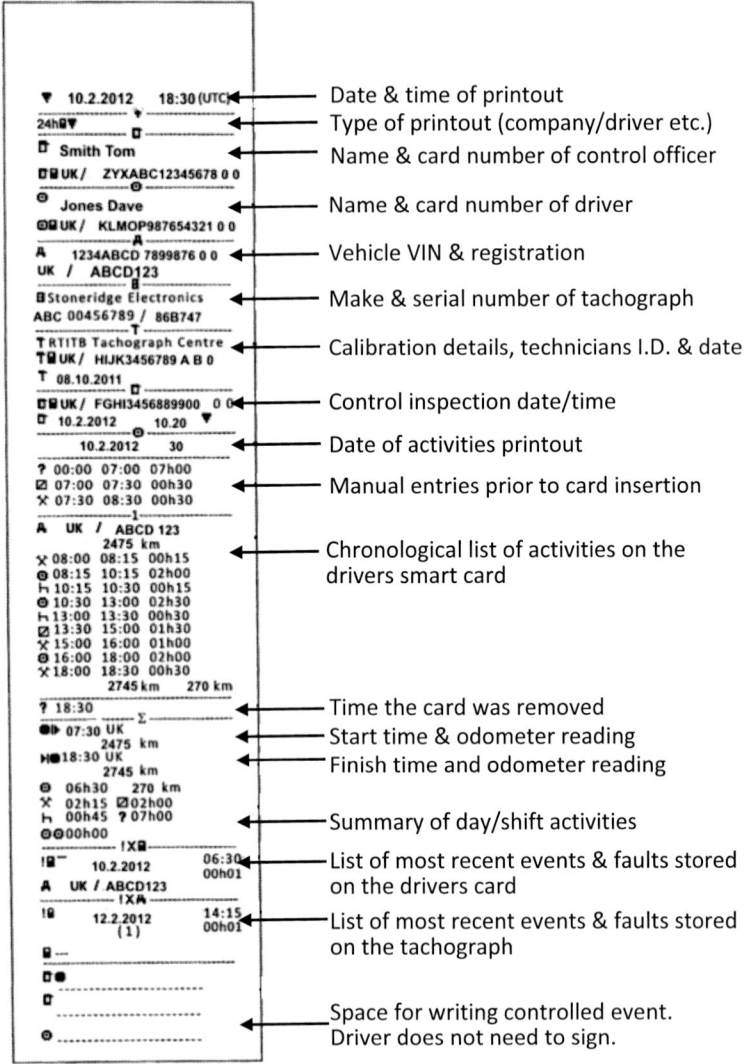

Information stored on the smart card includes:
- A photograph of the holder
- Driving Licence number
- Driver's signature
- Identification information
- Date of expiry

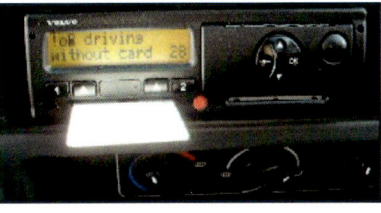

Types of Cards
There are four types of smart card in circulation. These are:
- The drivers card (for use only by that driver)
- The company card (for use by the operator)
- The workshop card (for use by the tachograph centre)
- Control card (for use by VOSA or the police)

Multi--Manned Vehicles
Where vehicles are crewed by more than one driver. Driver one must place his card in slot one and the driver who is not driving should place his card in slot two. When drivers change over they must also swap their cards over.

Downloading of Data
The downloading of data from drivers' smart cards should take place at regular intervals. Cards must have their data downloaded within a maximum period of 28 days, after which time the card will be full and any new data will start to be recorded over the old data. There is also a requirement for drivers to return records and printouts to their employer/contractor within 30 days. If a driver has lost or misplaced his or her card they must report it to the issuing authority (DVLA) and apply for a new card within 7 days.

The Keeping of Records
Drivers must retain records of their driving for the current day and the previous 28 days. Under the drivers' hours regulations, employers must retain records for one year. However, under the working time directive (WTD), employers must retain records of work for two years. So if no other record of the drivers working times are kept by the employer, they must keep the tachograph records for two years.

Questions on Tachographs

Q1. For how long must your employer keep your tachograph charts/records?
A. 1 year
B. 2 years
C. 18 months
D. 3 years

Q2. What must you record in the centre of an analogue tachograph before starting out with that vehicle?
A. Your name and vehicle registration
B. Your date of birth
C. The weight of the load you are carrying
D. The name of your employer

Q3. What information is recorded by the tachograph?
A. Weight of the load
B. Time, speed of the vehicle and distance travelled
C. How often you stop for fuel
D. How fast the engine is revving

Q4. For what period of time does a single analogue tachograph chart last?
A. 12 hours
B. 48 hours
C. 24 hours
D. 36 hours

Q5. How often will the analogue tachograph require checking?
A. 2 years
B. 1 year
C. 6 years
D. 4 years

Q6. Can drivers be prosecuted for speeding using the evidence from the tachograph chart alone?
A. Yes
B. No
C. Sometimes
D. Only if the vehicle is a coach

Q7. Can I use my friend's smart card in order to drive a vehicle fitted with a digital tachograph?
   A. Yes
   B. No
   C. Only if he gives you permission
   D. Only when he is not driving

Q8. If you don't put the smart card into a digital tachograph, will the tachograph still record any information?
   A. No
   B. Some information but not all
   C. Only the speed of the vehicle
   D. Yes

Q9. How often will my digital smart card need replacing?
   A. 24 hours
   B. 1 year
   C. 5 years
   D. 10 years

Q10. What do I do if I have lost my smart card or have had it stolen?
   A. Report it within 24 hours
   B. Don't report it; just apply for a new one
   C. Report it within 7 days and apply for a new one
   D. Do nothing just use a spare

Q11. Do I need to carry my smart card even if I only have an analogue tachograph in my cab?
   A. Yes
   B. No
   C. Only if you are going to drive a vehicle with a digital tachograph
   D. Only if you think you are going to be stopped

Q12. What do I do if I can't find my digital smart card?
   A. Report it missing to the police as soon as possible
   B. Ask your employer for a new one
   C. Borrow a friend's
   D. Report it missing to the DVLA within 7 days

Q13. If I drive vehicles using both types of tachograph, do I need to keep both my smart card and analogue discs for inspection?
   A. No - only the analogue discs
   B. Yes - both
   C. No - just the smart card
   D. Just the ones for this week

Q14. How often will I need to download the data on my smart card?
   A. Only when it's full
   B. Only when my employer asks for it
   C. Only if I get stopped by the police
   D. Once a week or before it becomes full at 28 days

Q15. Can I use my operator's (employer's) smart card if mine is lost?
   A. Yes you can use the operator's card
   B. Only operators can use the card to download information from the tachograph
   C. The operator's card can only be used in an emergency
   D. The operator's card can only be used when vehicles are out on test

UNDERSTANDING YOUR VEHICLE AND ITS CHARACTERISTICS

Before you drive your vehicle it is vital that you have a complete understanding of the vehicle, its characteristics and how it will perform. This is of even greater importance if it's your first time at driving that vehicle.

Safety is your main priority and you will need a clear understanding of the following.

You will need to familiarise yourself with all the controls and the seating position, making any adjustment necessary before driving away. Use your walk-around check to familiarise yourself with the vehicle.

Take a good hard look at your vehicle and familiarise yourself with its physical dimensions, size and weight. You will need to know and understand:

- Its **Length** when cornering and overtaking.
- Its **Height** when driving under bridges, trees and other possible obstructions.
- Its **Width** when passing through narrow gaps and openings.
- Its **Weight** both loaded and unloaded.

As the driver, it is your responsibility to understand your vehicle, its characteristics and performance. So you will also need to know and understand the following:

1. How long it will take for your vehicle to stop when loaded or unloaded.
2. How to control your vehicle at all times, particularly when travelling downhill.
3. How much time and distance you will need to overtake.
4. How much power and which gear to use when travelling uphill.
5. How to control your vehicle and make subtle and not violent changes to the controls.
6. How to attach and detach trailers safely (if required).
7. How your vehicle behaves when loaded and unloaded.

8. How to load and unload your vehicle safely.
9. How to distribute and secure the load.
10. How to operate all auxiliary equipment on the vehicle.
11. How your vehicle will behave in poor weather conditions such as high winds.

You need to understand your vehicle and all its idiosyncrasies and characteristics. If it's a new vehicle or not your usual vehicle you will need to take time to get to know and understand how it is going to behave.

Remember: - You are a professional driver. You are in control of your vehicle and safety is your priority.

Questions on understanding the characteristics of your vehicle

Q1. When driving any vehicle for the first time, you should…
   A. Just get the keys and go
   B. Check the vehicle over and spend some time getting to know the vehicle and making adjustments
   C. Ask another driver if he/she has driven this type of vehicle
   D. Just sit in the driver's seat and start the engine

Q2. When driving your vehicle your first priority is…
   A. To get to your destination as quickly as possible
   B. To put the biggest load you can on your vehicle
   C. To operate your vehicle as safely as possible
   D. To ensure your vehicle has enough fuel for the journey

Q3. You are required to drive your large vehicle through a busy town in the middle of the day, you should…
   A. Reduce your speed and be aware of other road users and pedestrians
   B. Keep a look out for the police
   C. Put your lights on to let others know you are there
   D. Sound your horn to let others know you are there

Q4. You are about to drive down a steep hill with a heavy load, you should…
   A. Give way to smaller vehicles
   B. Pick up speed to save time
   C. Change down through the gears to control your speed
   D. Keep your foot on the brake all the way down the hill

Q5. When driving with a heavy load you should always make subtle changes to the controls in order to...
   A. Show how good a driver you are
   B. Keep your speed up
   C. Keep full control of your vehicle
   D. Use less fuel

Q6. When your vehicle is fully loaded as opposed to being empty, it will...
   A. Take less time to stop
   B. Take the same time to stop
   C. Take less time to accelerate
   D. Take more time to stop

Q7. When connecting a trailer to your rigid vehicle, you should take account of...
   A. The additional length and weight of the vehicle
   B. The increased height of the vehicle
   C. The increased width of the vehicle
   D. The need to accelerate faster

Q8. When turning left at a busy 'T' junction with a long vehicle....
   A. Keep in tight to the near side corner
   B. To save time try and force your way out
   C. Check your near side mirror and move out into the road as far as possible before turning
   D. Drivers of large vehicles should never turn left at road junctions

Q9. When driving a high sided vehicle in strong cross winds...
   A. Put the radio on to listen to the weather forecast
   B. Try to complete your journey as quickly as possible
   C. Reduce your speed and be aware of gusts in open or exposed areas
   D. Keep you foot on the brake at all times

Q10. After loading your coach with heavy luggage, you should...
   A. Telephone you boss and tell him you're worn out
   B. Take account of the additional weight of your vehicle
   C. Take a break before driving
   D. Tell you passengers to bring less luggage next time

MORE EFFICIENT USE OF FUEL

As a driver the amount of fuel you use is directly related to the speed of your vehicle. The faster you go the more fuel you use. By keeping your vehicle in the same gear and reducing your speed just a little you can save on the amount of fuel you use.

The cost of running a modern commercial vehicle is significant and fuel plays a big part in that cost.

The table below shows the range of fuel consumed by a large vehicle travelling at different speeds in the same gear.

Speed (MPH)	Distance Travelled (Km)	Fuel use (Litres per hour)
Engine Idling	0	1.9
37	22.2	4.1
50	22.2	6.6
56	22.2	8.4

Source: IRTE.

Listed below are some of the ways of saving fuel.

1. A large vehicle travelling at 50 mph can use most of its power just to push a hole through the air. By reducing your speed you can save fuel.

2. By spotting and reacting to approaching hazards early you can take appropriate action to reduce fuel consumption in the following ways:

   i. By taking your foot off the accelerator early, you reduce the need for heavy braking, you will also save on fuel.
   ii. By keeping your vehicle moving (rolling) you will reduce the amount of energy needed to get it going again and will save fuel.

3. Avoid letting your vehicle idle for long periods. This simply uses up fuel unnecessarily.

4. Use fewer gear changes (block change). You don't have to use all the gears, particularly when empty. This will mean less work for you and a more fuel efficient vehicle.

5. Check your tyre pressures. Under inflated tyres will affect the rolling motion and the vehicle will use more fuel.

6. Check fuel lines for leaks and don't overfill the fuel tank particularly in warm weather as this can cause fuel leakage through expansion of the fuel. Make sure that the fuel cap is Fitted correctly after filling. Loss of fuel can lead to accidents and will increase fuel consumption.

7. Plan and, if possible, time your journey to avoid traffic. This will save fuel.

8. As far as possible try and use motorways and dual carriageways, as you tend to use higher gears on these roads, which means that you use less fuel for each mile travelled.

Questions on Fuel Efficiency

Q1. When driving the saving of fuel through more economic running of the vehicle can...
    A. Cost the company more in fuel costs
    B. Reduce fuel costs
    C. Increase the wear and tear of the vehicle
    D. Create greater risk of an accident

Q2. Tyres that are under inflated can...
    A. Increase the fuel consumption
    B. Reduce the fuel consumption
    C. Make the vehicle go faster
    D. Make the tyres last longer

Q3. Over filling the fuel tank in warm weather can...
    A. Improve the fuel economy of the vehicle
    B. Improve the running of the vehicle
    C. Cause fuel spillage due to expansion of the fuel
    D. Improve the efficiency of the engine

Q4. An engine idling for long periods can improve fuel efficiency...
    A. True as you are not revving the engine
    B. True because it's not making much noise
    C. Untrue as an engine idling for long periods uses up fuel
    D. It stops the engine from overheating

Q5. Letting the vehicle slow naturally without braking can improve fuel efficiency...
   A. True - as the brakes take energy out of the vehicle
   B. Untrue - as braking helps with more efficient driving
   C. The harder you brake the more fuel you save
   D. Braking has no influence on fuel efficiency

Q6. Keeping the vehicle rolling in traffic, as opposed to stop starting, can improve fuel efficiency...
   A. Untrue - stop starting is more efficient
   B. Makes no difference
   C. True - as starting from rest uses more energy and fuel
   D. Traffic saves fuel as you are going slower

Q7. Speeding helps with fuel efficiency as you get to your destination faster...
   A. This is true and you can save fuel this way
   B. Untrue as the faster you go the more fuel you use
   C. Makes no difference
   D. The faster the vehicle travels the more efficient it is

Q8. Can planning your journey help with saving fuel?
   A. No as you still have to get to your destination
   B. No because it just takes up more time
   C. Yes because you can avoid hazards en-route
   D. Only if you are carrying a heavy load

Q9. Does the type of road have any effect on the amount of fuel your vehicle uses...
   A. No effect at all
   B. Only a little effect on large vehicles with heavy loads
   C. Yes because hilly routes and traffic force drivers to use lower gears
   D. Yes because minor roads are better because there is usually less traffic

Q10. Can reducing your speed improve fuel efficiency?
   A. No as the faster you go the quicker you get there and the less fuel you use
   B. Yes a slower speed in high gears improves fuel efficiency
   C. Only when the vehicle is empty
   D. Only when going uphill

VEHICLE SAFETY AND WALK AROUND CHECKS

The safe operation of your vehicle is vital not only to you but also the public at large. Drivers therefore need to undertake daily checks of their vehicles in order to ensure that safety is a priority.

Attached to this section are two documents. They will help you identify all the checks that you should make on a daily basis. Should you find a problem, you will need to report it to your supervisor, employer or maintenance team. Your employers may have their own forms. Take a good look at the documents and familiarise yourself with them.

When reporting faults, you will need to apply a bit of common sense. Firstly, all faults need to be reported in writing. This helps to ensure that they get put right and not forgotten, which can happen even in the best workshops. If it's a more serious fault that affects the safety of the vehicle, report it immediately you notice it. More minor faults found during your shift can be reported at the end of your shift.

You should also be aware that if you don't carry out a check on your vehicle and it has had minor accident damage caused by the previous driver, you could end up getting the blame. As a driver you are responsible for your vehicle.

You should also know that the defect report form is a legal document and once the fault has been rectified by the engineer/mechanic it will need to be signed off by him or her. All maintenance records must then be kept available for inspection by VOSA for 15 months from the date on that record, unless it is a nil report, in which case it needs to be kept until the vehicle's next safety inspection.

Question on Walk-around Checks

Q1. How often should you carry out a walk-around check?
    A. Twice a week
    B. Every day after driving
    C. Only when driving a different vehicle
    D. Every day before driving

Q2. Whilst undertaking a check you find a deep cut in a tyre, should you?
   A. Report it on your return
   B. Let the maintenance team know immediately
   C. Report it but still take the vehicle out
   D. Not report it at all

Q3. It's far better to advise of faults in writing for the following reason.
   A. It will be on record and is less likely to be forgotten
   B. It makes you look good in front of the boss
   C. You don't have to talk to the mechanic
   D. It will show that you can write

Q4. Whilst driving, your vehicle develops a serious fault with the foot brake. You should do the following.
   A. Stop and report the fault immediately
   B. Report it as soon as you get back to the yard
   C. Wait until it gets worse
   D. Just use the secondary brake to get you home

Q5. The radio in your vehicle has stopped working, you should do the following.
   A. Refuse to take the vehicle out
   B. Report the fault in writing
   C. Ask another driver for assistance
   D. Hold your mobile to your ear to listen to music

Q6. The main reason for walk-about checks is to
   A. Find work for the mechanics
   B. Try and find a way of not working that day
   C. Ensure that the vehicle is safe and serviceable
   D. Give yourself something to do when not driving

Q7. You have reported a fault but still decide to take the vehicle out before it is repaired. As a result of the fault, the vehicle is involved in an accident.
   A. You are not to blame because you reported the fault
   B. The mechanic is at fault because he did not repair the fault quickly enough
   C. The responsibility lies with you as you decided to drive the vehicle knowing it had a fault
   D. Only the company is liable

Q8. The fuel gauge on your vehicle is playing up away from home. What should you do?
   A. Report it and get it repaired
   B. Just get a new one fitted at the nearest garage
   C. Record the mileage and try and keep the tank full until you get home
   D. Do nothing and hope for the best

Q9. On returning to the yard, you discover that there is a slow puncture in one of your tyres. What should you do?
   A. Report it immediately
   B. Leave it for the next driver to sort out
   C. Hope it stops going down
   D. Report it the next day when you get to work

Q10. As a driver you could be prosecuted for driving an unsafe vehicle.
   A. This is true
   B. No - the company has full responsibility
   C. Only the mechanics have that responsibility
   D. Only when driving a loaded vehicle

Driver's Walk around Check list (PCV)

Vehicle Registration _____ Mileage _____ Date _____ Sign _____

From inside the vehicle	Detail of Check	Tick or Comment
Seats & Seatbelts	Check that the driver's seat and seatbelt are adjusted and function correctly.	
Mirrors & Glass	Check that all mirrors & windows are clean, free of cracks or damage & adjusted correctly.	
Horn	Check the operation of the horn.	
Brakes	Check air pressure gauges and warning lights are working. Check the operation of all brakes including those of trailers and semi-trailers. Check the operation of the brake pedal, any free play and anti-slip provision.	
Steering	Check for excessive play in steering gearbox or linkage.	
Wipers & Washers	Check the function, efficiency and operation of wipers and washers.	
Exhaust Smoke	Check for excessive exhaust smoke.	
Controls & Dashboard Operation	Check that all controls and warning lights operate correctly when the engine is switched on.	
Heating Ventilation & Air Conditioning.	Check the operation of forced air systems and de-mister equipment.	

From inside the vehicle	Detail of Check	Tick or Comment
Doors and Exits	Ensure that all doors and exits are fully operational, secure when shut and all exit signs are in place. Check the security and operation of all windows and roof hatches.	
First Aid Kit	Check that the first aid kit is present (if applicable), in good condition and easily accessible.	
Fire Extinguisher	Ensure that the fire extinguisher is easily accessible, in good order and is the correct type – water or foam.	

From outside the vehicle	Detail of Check	Tick or Comment
Lights, Indicators and Reflectors	Check that all lights, indicators and reflectors are clean and are operating correctly. (Get assistance with brake lights).	
Wheels and Tyres	Check that all tyres are legal with no cuts or damage. Ensure that there is no debris trapped between twin wheels. Check for damage and the security of all wheels including the spare.	
Fuel, Water and Oil Leaks	Check for visible oil, water or fuel leaks from engine, transmission, fuel lines or tanks and steering components.	
Battery Security	Check that all batteries are secure, free of leaks and corrosion. (Only if accessible).	
Brake and Electrical lines	Listen for air leaks and look for a drop in air pressure.	
Security of Vehicle Body work	Check the security of all body parts and luggage compartments. Ensure that all fastening devices, including locks, are in good operational order.	

VEHICLE DEFECT REPORT

Name (in capitals)		Vehicle mileage	
Reg no.		Fleet or trailer no.	

Daily Checklist

Vehicle body/ exterior	All doors, compartments and locks	Heating and ventilation	
Mirrors	All seats and seatbelts	Vehicle interior	
Glass/reflectors	Warning lights and buzzers	Engine exhaust smoke	
Tyres / wheels	Vehicle lights	Steering	
Vehicle load (if applicable)	Indicators	Brakes	
Fuel / oil leaks	Wiper / washers	Battery / connections / security (if accessible)	
Horn	Brake lines (if accessible)	Fire extinguisher	
Coupling or trailer security (if applicable)	First aid kit		
Faults / defects (if found)		Faults / defects rectified	
Drivers signature		Faults / defects rectified by:	
Date:		Date:	

WHEELS AND TYRES

As a driver it is your responsibility to check your wheels and tyres on a regular basis. They should be checked as part of your daily checks. If you change vehicles you need to check the wheels and tyres before you leave the yard.

So what should you be checking?
- Check that the tyres are inflated correctly. Under inflated tyres increase the wear rate of the tyres and also increase fuel consumption. The tyre pressure should be checked when the tyre is cold because the pressure increases as the tyre heats up.
- Check that they do not have any deep cuts or blisters. Should you find any cuts or blisters in a tyre, you need to report it immediately. Do not try to put more air in a tyre that has damage. Most commercial vehicle tyres are inflated to around 100 Psi or more. In a large tyre that sort of pressure adds up to be well over 30 tonnes across the whole tyre and it will kill if there is a blow-out when you are stood next to it. Blow-outs from commercial tyres are commonplace. As drivers you know that when travelling down a motorway you can see tyre debris every couple of miles.

Responsibility
As a driver you need to be aware that it is you who are responsible for the safe operation of your vehicle. If stopped on the road by the police or VOSA, your vehicle could be subject to a prohibition notice and you could receive a fixed penalty notice if your vehicle is found to be a danger to other road users.

You should also:
- Check the tread depth, and look for uneven wear on the tyre. There should be a minimum of 1.00 mm of tread across the surface of the tyre. Lighter vehicles under 3.5 tonnes MPW require 1.60 mm of tread.

- If your vehicle is fitted with double wheels check that there is no debris stuck between the wheels especially if the vehicle has been off road.
- Some companies do not allow drivers to change their own wheels and tyres due to health and safety regulations and company policy. The dangers are such that only experienced engineers or companies should undertake the job. It's not just the risk when jacking up the vehicle but also the risk from other traffic particularly on motorways. Drivers need to establish what their company policy is on changing wheels.

Wheel Safety

In a recent report by the Department of Transport the importance of correct wheel fixing and maintenance procedures was highlighted. Research showed that up to 400 wheel detachment incidents occur each year of which up to 27 result in injury, accidents and up to 7 in fatal accidents. The report recommended the use of wheel nut retention devices or movement indicators, as these devices were shown to be effective.

So when checking your wheels you should look for loose or missing wheel nuts. Remember, all wheel nuts need to be tightened using the correct procedure and to the correct torque as recommended by the manufacturers of the vehicle.

Look for damage to the wheels such as cracks or splits.

Wheel Nut Indicators
This type of wheel nut indicator gives a clear visual indication of nut rotation or nut absence to drivers and mechanics as part of their routine vehicle inspection. However they will not stop the wheel nuts from coming loose.

Questions on Wheels and Tyres

Q1. When checking and tightening wheel nuts you should...
    A. Make sure they are as tight as you can make them
    B. Only tighten the ones that look loose
    C. Ensure that all the nuts are tightened to the correct setting with a torque wrench
    D. Just check one or two and the rest should be fine

Q2. Tyre pressures should be checked ...
    A. At least every month
    B. At least every week
    C. Only when carrying heavy loads
    D. Only if they look low

Q3. Wheel nut indicators are used...
    A. To help the driver see if any of the wheel nuts are working loose
    B. To make the wheels look good by adding colour
    C. So that VOSA and Police can check the wheel nuts
    D. So that you know that the wheels nuts have been tightened up

Q4. A large commercial tyre that has been inflated to 100 Psi will have approximately...
    A. Ten (10) Tonnes of pressure across the whole tyre
    B. Only five (5) Tonnes of pressure across the whole tyre
    C. One hundred (100) Tonnes of pressure across the whole tyre
    D. Over thirty (30) Tonnes of pressure across the whole tyre

Q5. The minimum legal tread depth on a large commercial vehicle is...
    A. 2.00 mm
    B. 1.60 mm
    C. 1.50 mm
    D. 1.00 mm

Q6. Drivers should undertake a visual inspection of their tyres every...
    A. Week
    B. Month
    C. Two weeks
    D. Day

Q7. You notice that the double drive wheels on your vehicle are kissing. This is an indication...
   A. Of over inflation of the tyres
   B. Of under inflation of the tyres
   C. That the tyres need changing
   D. All is fine and there is no need to do anything

Q8. A 10 Psi drop in pressure of a large commercial tyre can result in...
   A. An increase in fuel efficiency
   B. The vehicle doing more miles to the gallon/litre
   C. The driver losing control
   D. A drop in fuel efficiency

Q9. Whilst undertaking your daily safety check you discover a deep cut in a tyre. You should ...
   A. Get a jack and change the wheel
   B. Report it as soon as possible
   C. Report it after your journey
   D. Leave it for the next driver

Q10. As part of your daily check you find debris between your double drive wheels. What should you do?
   A. You should try and remove it, if you can't report it
   B. Ignore it
   C. Let the tyres down and hope it drops out
   D. Try reversing in order to move it

DIMENSIONS AND WEIGHTS OF VEHICLES

The rules relating to the size and weights of vehicles in the EU are very complex and can differ between each country. So in this section we will deal with the regulations as applied to the UK. These rules come under the construction and use regulations of the Road Traffic Act.

Below are listed the maximum weights for each type of vehicle layout. However, to be sure, you should check the vehicle plate which will show the maximum design or gross weight, the maximum weight for each axle and the maximum permissible weight if towing a trailer.

Maximum Gross Vehicle Weights	
Two-axle vehicles	18 tonnes
Three-axle vehicles	26 tonnes
Four-axle vehicles or combinations	32 tonnes
However, in order to carry the maximum permissible weight in each category vehicles must meet lower emission standards and have a drive axle limit of 11.5 tonnes.	

Maximum Axel Weights	
Single driven axle	11.500 kg
Single non-driven axle	10 000 kg
Tandem driven axles	19 000 kg

Vehicle Lengths
The maximum length is the distance from the foremost part of the vehicle to the rearmost part of the vehicle. The maximum width is distance between the most protruding parts on each side excluding mirrors.

Vehicle	Maximum length
Rigid Vehicles	LGVs = 12 metres PCVs = 15 metres
Articulated Vehicles	15.5 metres (However, this can be increased to 16.5 metres under certain conditions.)
Vehicle and drawbar combinations	18.75 metres

N.B. It should be noted that there are additional regulations that apply in all cases to items such as turning circles, drawbar length and load space.

Vehicle Widths
The maximum width for most vehicles is 2.55 metres. The maximum width of a trailer is 2.55 metres provided that the towing vehicle's maximum permissible weight is more than 3.5 tonnes.

Vehicle Heights
The EU set a maximum height of 4 metres. However, this rule does not apply in the UK where there is no maximum height limit for goods vehicles. That said, common sense must prevail, so if you load a vehicle to say six metres it will not go under bridges and will bring down cables crossing the road. There is however a maximum height for buses, this is 4.57 metres or 15 ft.

Care should be taken when driving under arched bridges as the height of the bridge will be lower on the near side than the middle of the bridge. Each year there are hundreds of incidents involving vehicles hitting bridges in the UK. Drivers need also to be aware of other dangers such as branches of overhanging trees, filling station canopies and problems caused by road cambers.

The minimum height for motorway bridges is 16 ft or 4.92 metres.

It is a regulation that all vehicles that have an overall height of more than 3 metres must have a visual display in the cab of the vehicle, showing the actual height of the vehicle and its load.

Questions on Vehicle Weights and Dimensions

Q1. If a vehicle of more than three metres in height is being operated it must
   A. Not travel off motorways
   B. Not be driven at more than 50 mph
   C. Have a warning sign in the cab advising the driver of the vehicles height
   D. Not be reversed

Q2. The maximum weight of a two-axle vehicle is...
   A. 16 tonnes
   B. 22 tonnes
   C. 17 tonnes
   D. 18 tonnes

Q3. The maximum width of a bus is 2.55 meters.
   A. This includes the mirrors
   B. This does not include the mirrors
   C. Includes the mirrors but only if they are retractable
   D. Includes the mirrors but only if the mirrors are more than 3 metres from ground level

Q4. The maximum length of any rigid bus or coach is...
   A. 10 metres
   B. 15 metres
   C. 18 metres
   D. 12.5 metres

Q5. The maximum height of a bus used on UK roads is...
   A. 4 metres
   B. 4.57 metres
   C. 3.8 metres
   D. No legal maximum height limit

Q6. The maximum width on trailers carrying general goods being towed by vehicles over 3.5 tonnes is...
   A. 3 metres
   B. 2.3 metres
   C. 2.6 metres
   D. 2.55 metres

Q7. The maximum length of a combination of vehicle and trailer is...
   A. 15.5 metres
   B. 18 metres
   C. 18.75 metres
   D. 18.5 metres

Q8. When driving on an unfamiliar route the driver should take extra care when approaching overhanging tree branches, this is because...
   A. They may restrict the light
   B. They may be a rare or uncommon species
   C. They may be too low for their vehicle to pass under safely
   D. A danger to other traffic

Q9. The overloading of an axle can result in the driver being fined a maximum of...
   A. £1,000
   B. £500
   C. £5,000
   D. £250

Q10. If towing a trailer with a vehicle weighing more than 3.5 tonnes, the maximum width of the trailer is...
   A. 2.5 metres
   B. 2.9 metres
   C. 2.3 metres
   D. 2.55 metres

TRANSMISSION SYSTEMS

All vehicles require a transmission system. The transmission is used to undertake a number of functions. This is because any engine has a limited amount of power; it can only rotate at a limited range of speeds and can rotate in only one direction.

The transmission is used to overcome these problems by multiplying the engine power (torque) allowing for a range of road speeds, changing the direction that the wheels rotate (forwards and reverse) and to be able to disconnect the engine from the road wheels whilst it is still running. (By placing the gearbox in neutral.)

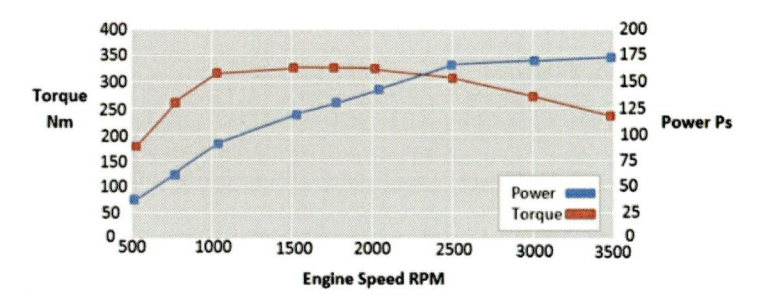

As stated above, engines have a limited amount of power. The proof of this is that you can never start from rest in top gear; there simply isn't enough power in the engine. We need to be able to multiply that power by selecting a lower gear in the gearbox (usually 1st). Lower gears allow us to improve the engine torque but in doing so we sacrifice speed. The lower the gear the less speed but more power (torque).

It is important to have some empathy with the vehicle when driving. Try and treat it with a little respect. There is no need to push the engine revs into the red before changing gear. You are not driving a Ferrari on the BBC's Top Gear. Most engines perform best (produce the most power & 'torque') at well below their maximum revs. (See graph above).

Equally don't let the engine labour in any one gear.

Types of Gearbox
There are many types of gearbox in use. These range from manual, semi-automatic to fully automatic. In addition drivers may also find that they may have a two speed rear axle giving additional multiplication of the gears in the gearbox. Whilst all gearboxes do more or less the same job, they operate in different ways.

Modern Gearbox
Old style crash gearboxes have for many years been replaced with more efficient synchromesh gearboxes. The need to double-declutch when changing gear has been passed to the history books. Most modern synchromesh gearboxes have six or more gears with either a splitter box (high and low ratios on the gearbox) or a two speed drive axle. This will give a possibility of twelve or more forward gear ratios. Generally the splitter ranges (either in the gearbox or axle) are changed electronically via a switch on the gear lever.

Fully Automatic Gearbox
Many modern gearboxes are fully automatic. Whilst this may reduce some of the skill of being a driver, it also reduces some of the stress and fatigue on the driver. A trip from the West Country to London could see the driver changing gear up to 300 times. A fully automatic transmission removes the need for a clutch and only requires the driver to decide if he/she wants to go forwards or backwards. Gears are selected automatically, based upon engine and road speed of the vehicle.

Latest Technology
Most manufacturers are now producing some really sophisticated and technological advanced transmission systems which read the road ahead anticipating changes in the topography and change gear automatically in order to deliver the best possible fuel consumption figures for that vehicle and its load.

Questions on Transmission Systems

Q1. When selecting a high gear you are...
    A. Increasing the torque at the road wheels
    B. Decreasing the speed of the vehicle
    C. Decreasing the torque at the road wheels
    D. Making no change to either speed or torque

Q2. When driving a vehicle with a manual gearbox the driver can...
   A. Only change up one gear at a time
   B. Only change down one gear at a time
   C. Block gear change by missing gears when changing up or down
   D. Only block change when changing down gears

Q3. When selecting a gear in a manual synchromesh gearbox the driver should...
   A. Always double de-clutch
   B. Change down a gear before changing up
   C. Only press the clutch down once
   D. Wait until the rev-counter is in the red

Q4. Fully automatic gearboxes will...
   A. Reduce driver fatigue
   B. Increase driver fatigue
   C. Make no difference to the stress on the driver
   D. Make the vehicle go faster

Q5. When travelling down a steep downhill gradient and when using a manual gearbox the driver should...
   A. Take the vehicle out of gear
   B. Select a low gear at the top
   C. Change up a gear
   D. Switch the engine off to save fuel

Q6. One of the main functions of the gearbox is to...
   A. Make the engine work harder
   B. Make more work for the driver
   C. Help reduce the work load on the clutch
   D. Multiply the engine torque

Q7. Having identified a hazard, by thinking ahead and changing gear early, the driver can
   A. Improve the fuel economy of the vehicle
   B. Increase the speed of the vehicle
   C. Increase the load that the vehicle can carry
   D. Ensure that the rev-counter always stays in the red

Q8. The maximum speed that a vehicle without a speed limiter can reach is determined by...
   A. The gear the driver has selected
   B. The way in which the gear was changed
   C. The size of the gearbox
   D. How many axles the vehicle has

Q9. Allowing the engine to labour in high gears may...
   A. Help the engine's performance
   B. Increase the engine's power output
   C. Allow the vehicle to go faster
   D. Cause damage to the engine

Q10. The amount of torque that a vehicle can deliver is determined by...
   A. How fast the driver can change gear
   B. Which gear the driver has selected
   C. How heavy the load is
   D. How fast the driver can travel downhill

COMMERCIAL VEHICLE BRAKING SYSTEMS

Using your Brakes
When using your brakes you are taking energy out of the vehicle. Very often simply by being observant and with some forward planning you can avoid any heavy braking and in many cases avoid the need to use your brakes at all. How many times have you been in traffic and seen the driver of the vehicle in front using their brakes every few seconds because they are tailgating the vehicle in front of them? This type of poor driving uses up fuel and causes wear and tear on the vehicle.

By being a bit more focused and planning ahead you can avoid heavy braking and in doing so reduce the risk of an accident. If you have to brake hard it is easy to lose control of your vehicle, particularly in winter if road conditions are poor. So think ahead and give yourself time. Remember the 2 second rule. In winter you may need to extend the rule to 8 or even 10 seconds, if conditions are really bad.

Braking System Types
- **Hydraulic Brakes**

    Smaller, lighter commercial vehicles tend to use a tandem hydraulic brake system which is supported with a vacuum servo. In addition the vehicle will have a mechanical hand or parking brake system. The systems used are usually similar to those used on large cars.

- **ABS and Electronic Systems**

    Many vehicles are now fitted with ABS (This stands for Antiblockiersystem and is a German word for Anti locking brakes). The systems which can be quite sophisticated are there to support the driver. The vehicle's braking system is at its most efficient when the wheels/tyres are held on the verge of skidding.

Once the wheels/tyres start to skid then two things happen:

(1) The vehicle does not slow as fast. (2) The driver can lose control. In the past drivers used to 'pump' the brake pedal. This is called 'Cadence Braking'. With ABS systems, a computer and sensors do much the same thing only much faster, holding all the wheels/tyres at a point just before they start to skid, thus giving the maximum braking efficiency and allowing the driver to keep control.

- **Air Brakes**
As vehicles become heavier so more pressure needs to be applied to the brakes to slow or stop them. In order to achieve that pressure, air is used. The pressure is created using a compressor. Most air systems work on a pressure of around 100 Psi. There are generally three systems in operation on each vehicle.

1. The service brake. This is usually the foot brake and the main brake. It can also have ABS (see above) incorporated in the system.
2. The secondary brake. This is the emergency or safety brake should the main brake fail in some way.
3. Parking Brake. This is usually a hand brake and used when the vehicle has stopped.

Brake Retarders
- **Exhaust Brake Function**
This type of brake is simply a large valve in the engine's exhaust stream (usually near the exhaust manifold.) When the valve is closed, most of the exhaust gas is trapped inside the cylinders. The engine continues to function as it would normally, but gets very little fuel, just enough to idle. The exploding fuel's gases have nowhere to go so they push back against the piston and restrict the engine rotation. The feel from the driver's perspective is like changing down a couple of gears and using the engine as a brake.

- **Electronic Retarders**
A Retarder is an independent brake system and works by applying a current from the vehicle's batteries to a set of electro magnets which slows the rotation of the transmission.

The Retarder works wear-free. It can eliminate 90% of all brake application. Thus the normal foot brake remains cold and is available for emergencies with the full braking power and no fading with hot brake linings. If used correctly significant cost savings can be made by reducing the down time of the vehicle, as the brake linings last up to 8 x longer, Downhill, higher average speeds can be achieved whilst safety is increased, and passenger comfort improved.

Questions on vehicle braking systems

Q1. Whilst driving you feel that the brakes on your vehicle are not working properly. What should you do?
   A. Carry on and report it when you get back to the yard
   B. Carry on and leave it to the next driver
   C. Stop as soon as possible and phone for help
   D. Speed up to get back to the yard faster

Q2. When completing your daily check you hear an air leak on your vehicle. Do you?
   A. Report it immediately
   B. Pretend you haven't heard it
   C. Report it on your return
   D. Check to see that the air pressure is OK. If so, drive on

Q3. Whilst driving on the motorway you find yourself close up behind another vehicle. You should...
   A. Flash the other driver to get out of the way
   B. Try and get closer so as to be more aerodynamic
   C. Drop back in case he/she stops quickly
   D. Try and find a little more speed to overtake

Q4. When driving down a long steep incline. You should...
   A. Keep your foot on the brake at all times
   B. Select a low gear at the top and let the engine hold you back
   C. Pick up as much speed as possible
   D. Make no change to the way you drive

Q5. After connecting a trailer to your vehicle, you should...
   A. Just drive off
   B. Take a break
   C. Phone the office
   D. Check the brake line connections, vehicle lights, trailer and any load

Q6. Your vehicle is fitted with an electronic retarder. You should ...
   A. Use it all the time instead of the foot brake
   B. Use it only when in heavy traffic
   C. Use it to control speed when travelling down long gradients
   D. Don't use it at all as it slows you down

Q7. After connecting a trailer to your vehicle, you find that the lights are not working. You should...
   A. Check all the bulbs
   B. Drive off as it's daylight
   C. Phone for help
   D. Check the connections and fuses

Q8. When driving a minibus, you notice a sudden and excessive amount of travel (movement) on the foot brake. You should...
   A. Pull over as soon as possible and report it immediately
   B. Use the handbrake instead
   C. Tell someone in the yard when you get back
   D. Leave it for the next driver

Q9. When driving a full coach you notice that you are losing air pressure. You should...
   A. Simply ignore it
   B. Immediately use mobile to let the boss know
   C. Drive to the nearest garage using just the hand brake
   D. Pull over at the nearest safe place and telephone for help

Q10. When parked on a steep incline you notice that the parking brake is not holding firm. You should...
   A. Use the exhaust brake if fitted
   B. Jump out quickly and put a brick behind the wheel
   C. Pick up the phone and ring your boss
   D. Hold the vehicle on the foot brake and move it off the incline. Then report it

UNDERSTANDING THE IMPORTANCE OF HEALTH AND WELL BEING WHEN DRIVING

In order to drive a commercial vehicle as every driver knows you need to be relatively fit. Regular health checks or assessments are part of the job. In addition to the initial medical at the time you first started driving, drivers need to have further medicals every 5 years from the age of 45 years and every year from the age of 65 years. If you work regular nights your employer is required by law to ensure that you have annual health checks, are fit to drive and have no underlying medical condition that will affect your ability to do the job. You will know from your own experience that the medicals are quite stringent and if you pay for them yourself expensive. However, if you do pay for it yourself, you are allowed to claim this expense against you income tax. As the medical examinations can be up to 5 years apart, if a driver develops a condition that may affect their ability to drive between check-ups they will need to notify the DVLA at Swansea about the condition. If you have concerns, talk to your Doctor.

Eyesight/Vision
In order to be able to drive any vehicle there is a minimum standard of sight required. You must be able to read a number plate at a distance of 20.5 metres or 67 ft. If you need glasses or contact lenses in order to be able to read the number plate at that distance you must use them when driving. The police now have the power to make a driver undergo an eyesight test.

All new professional drivers are required to undergo an eyesight test as part of a medical examination. On reaching the age of 45 years and every 5 years thereafter drivers are required to undergo a full medical which will include an eye test. Commercial vehicle drivers who work nights will need to have an annual health check.

Stress
It's no secret that driving aggressively and reacting to other drivers on the road increases your stress levels. Government figures show

that up to 60% of all absences from work are caused by stress. That's about 270,000 people in the UK alone that take time off every day because of stress. We know that driving causes stress and forces us to take time off work. We know that stress makes us depressed and affects our relationships, our social life and our health.

One cause of stress is the feeling of lack of control; feelings of helplessness. An example of this where you have been given responsibility but none of the authority to deal with issues. There is a direct connection between lack of control and stress; this can lead to other symptoms such as fatigue. One study of Danish bus drivers found that when people took more control of their lives, cardiovascular diseases decreased. Bus drivers are at a higher than normal risk for these diseases for a number of reasons. These can include stress, diet and lack of exercise.

Dealing with Stress When Driving
We know that driving can cause a great deal of stress. Angry clients, numerous deadlines, the incompetence of some people, demanding bosses, waiting to be loaded or unloaded - the list just goes on and on.

But what many people do not realise is that much of this stress can start before you even begin work. In many cases, stress begins at home. Worrying about your partner, the kids and the bills can cause a real headache. Then you arrive at work and start driving and the stress gets worse. Heavy traffic, street noise and bad driving by others all contribute to a negative, tired, and irritable day.

We know that driving causes stress and forces us to take time off work. We know that stress makes us depressed and affects our relationships, our social life and our health. How can we reduce this? Here are some ways to help deal with stress on the road.

We know it's not easy when you are on a timetable, but try and allow extra time on the road. Cutting it too close from the start will increase blood pressure and cause a lot of stress because you need to drive faster and try and dodge traffic. It also increases the chances of being involved in road accidents. If you can, leaving early gives you more time to relax a little on the road. Where possible, driving other than at peak time will also help.

You're the driver, take control. It's a fact that when you are in control of a situation you feel less stressful. The Danish study above showed that this works.

You may also want to install other equipment that can help calm you down. If possible and if you can change your route, use an on-board navigation system to find alternative routes on major roads to avoid traffic hot spots. However, probably the most helpful way of reducing stress is to talk to a good friend. Be open and tell them what's bothering you. Also, listen to them as they probably have the same issues. That way you know you are not on your own.

Handling conflict
Dealing with aggression and conflict is probably one of the most stressful situations most people have to deal with. Unfortunately some jobs get more than their fair share of angry people. Customer complaints departments, call centres and bus drivers are amongst the top ones.

1. Don't look for conflict.
2. If someone is being aggressive keep calm.
3. Avoid physical contact.
4. Take a non-aggressive posture. (Don't invade their space).
5. Ask how you can help with any problem they may have.
6. Try not to be negative.
7. If possible, try and inject some humour into the situation.
8. Control your breathing.
9. Speak softly and don't raise your voice.
10. If all else fails walk away.

Road Rage
The UK has one of the worst records for road rage incidents in the world, with men being more likely to commit an aggressive act than women. Most people feel that when sitting in their cars they can behave badly because they have anonymity i.e. nobody knows who they are. This is not the same for professional drivers with their large sign written vehicles. The fact is that motoring is about dealing with other people and most importantly yourself. If you lose your own self control you have lost the argument. Drivers need to be self aware. Dealing with and reacting to bad driving is another form of conflict that can be a major cause of stress.

Driving after drinking alcohol or taking drugs
According to DfT figures, one in seven of all deaths on Britain's roads

is alcohol-related. This means that in this country, approximately 5 people are killed each week by drunk drivers.

Despite high-profile government campaigns targeting drivers, and ever harsher penalties imposed on those who drink alcohol before driving, thousands of people continue to flout the law, putting their, and the lives of others, at risk. It is well known that alcohol depresses the body's physical and mental responses and our ability to make reasoned and sensible judgements. Anyone who has ever consumed a lot of alcohol will tell you of their inability to reason, think or undertake a range of sometimes quite simple activities. Driving any vehicle after drinking alcohol even if you are an experienced driver will place you and other road users at risk.

Drink-driving: Limits and Penalties
So how much alcohol can put you over the legal limit? The answer is that we simply don't really know. There are so many variables; the weight and condition of the person. Have they had a meal? Are they taking any medication and many other factors which can affect their condition?

The UK drink-drive limit when measured is:
- Breath is 35 micrograms of alcohol per 100 millilitres of breath.
- Blood is 80 milligrams of alcohol per 100 millilitres of blood.
- Urine is 107 milligrams of alcohol per 100 millilitres urine.

If a driver is found to be above the legal limit (above), he or she can be jailed for up to six months, be fined up to £5,000; and be disqualified from driving for up to a year. In most cases, disqualification is automatic. You should be aware, that even if you are just marginally over the limit, you will still lose your licence for 12 months. If it's a second drink driving offence within ten years you will receive a 3 year ban. In the far more serious case of causing death whilst driving under the influence of alcohol, the offender can receive a maximum sentence of 14 years in prison, an unlimited fine, and an automatic two-year driving ban.

Many drivers fail to understand how long alcohol can stay in the body and have found to their cost that they have failed a breath test when driving the following morning after having an evening out.

Drugs and Driving
Many people when they hear or see the word drugs think generally of illicit substances. The truth is that we nearly all take some form of drug/medication when we are ill or have a medical problem. These can in some cases be as dangerous as alcohol.

Drivers need to be aware that there are many drugs that you can buy over the counter without a prescription that can affect your ability to drive. The picture shows just some of the more common medicines that can make you sleepy and impact on your ability to concentrate.

If you plan to take medication for anything read the label first!

The police are currently testing a number of devices to measure the type and amount of substances that drivers may have consumed.

Smoking
With changes in the law it is now illegal to smoke when on board your vehicle, even if you are not driving. This is because it is your place of work. Drivers who smoke whilst in their vehicle can attract a fine of up to £200. Smoking can also damage your heath. A recent survey shows that people are now living longer. One reason for this is that the number of people who smoke is going down. This increase in longevity (the number of years that we live) is set to increase. The average life expectancy for a man is now 80 years and a woman 83 years, although men are catching women up. The average life expectancy for smokers is significantly less at around 65 years. So do the maths, if you smoke you could be giving away years of your life.

Driver Fatigue
Driver fatigue is a very dangerous condition created when a person is suffering symptoms of fatigue while driving, often resulting in an hypnotic state. This effect, especially during night time, can result in drivers either falling asleep at the wheel or being so exhausted they make serious and sometimes fatal driving errors.

The early hours of the morning and the middle of the afternoon are the peak times for fatigue accidents. Also long journeys on monotonous roads, such as motorways, are the most likely to result in a driver falling asleep. Sunlight signals our bodies when to be awake, but when deprived of any natural light at night, we will feel a surge of fatigue. We feel fatigue to a lesser extent, in the middle of the afternoon, particularly after a heavy meal or alcohol. The latest research shows that the grogginess felt right after you wake up can also be dangerous.

Figures from DfT show that 300 people a year die from sleep related road accidents in the UK. Proper diet and rest may reduce the possibility of sleep apnoea (falling asleep). Statistics from DfT show that 40% of sleep related road traffic accidents involve commercial vehicle drivers. 80,000 or 1 in 6 of all commercial drivers have some experience of falling asleep at the wheel.

Healthy Eating
Proper nutrition and physical conditioning are important influences on the effects of fatigue for commercial vehicle drivers. Proper eating during the day significantly influences your alertness.

Reasonable physical conditioning and weight control also helps increase stamina and will enable you to feel better and simply add quality to your life.

As part of a healthy diet specialist nutritionists are telling us it is important that we eat at least five portions of fruit and vegetables every day.

They are also telling us that we need to cut down on red meat and foods that contain saturated fat.

Being overweight can lead to other health issues such as heart problems, high blood pressure and type 2 diabetes.

Questions on the importance of health and well being when driving

Q1. Stress can cause illness and in many cases force people to take time off work...
   A. Not true
   B. Only rarely
   C. This is true. Stress affects many people
   D. Only older people are affected by stress

Q2. One way of successfully dealing with aggression is to...
   A. Hammer the other guy before he starts
   B. Speak softly and don't raise your voice
   C. Shout back at him
   D. Ring the boss and ask for help

Q3. Driving a large vehicle in heavy traffic can be stressful...
   A. Only in London traffic
   B. Only during the rush hour
   C. True, the driving of any vehicle in traffic can be stressful
   D. Only when trying to meet deadlines

Q4. Some standard over the counter medications can cause drowsiness...
   A. Not true
   B. Only prescribed medication can cause drowsiness
   C. This is true and you should always read the instructions
   D. Only at night

Q5. According to the DfT the percentage of sleep related road deaths involving commercial vehicle drivers is...
   A. 10%
   B. 20%
   C. 30%
   D. 40%

Q6. Road rage incidents are very common in the UK...
   A. True. The UK has poor reputation for road rage
   B. Not true - many other countries are far worse
   C. True - but women are far worse than men
   D. Not true as I don't know anyone who gets angry when driving

Q7. According to the DfT how many people die from sleep related road traffic accidents in the UK?
   A. 100
   B. 200
   C. 300
   D. 400

Q8. The maximum penalty for drink driving when not involved in an accident is...
   A. Jailed for up to six months, fined up to £5,000, and disqualified from driving for up to a year
   B. Fined up to £2,000, and disqualified from driving for up to a year
   C. Fined up to £5,000, and disqualified from driving for up to two years
   D. Jailed for up to 2 years, fined up to £5,000, and disqualified from driving for up to a year

Q9. One way of reducing stress whilst driving is to...
   A. Drink and drive
   B. Listen to music
   C. Use aggression to all other road users
   D. Tell the boss what you think of him

Q10. The maximum breath alcohol limit without being over the drink drive limit is...
   A. 25 micrograms of breath per 100 millilitres of alcohol
   B. 45 micrograms of breath per 100 millilitres of alcohol
   C. 30 micrograms of alcohol per 100 millilitres of breath
   D. 35 micrograms of alcohol per 100 millilitres of breath

BASIC FIRST AID TECHNIQUES

Knowing what to do can make the difference to a person's recovery and you could even save their life.

First Aid Kit
It is advisable to have the following:

- First Aid leaflet
- Disposable gloves (non-latex)
- Range of dressings
- Range of bandages (including triangular bandages)
- Cleansing wipes
- Plasters
- Tape
- Scissors

Buses and coaches must have a first aid kit unless they are a local service.

What to do in an emergency
First of all you must assess the casualty using **the primary survey**, a quick logical sequence going through actions in a priority order.

D **D**anger
R **R**esponse
A **A**irway
B **B**reathing
C **C**irculation

DANGER
ASSESS the situation:
- Are there any risks to you or the casualty?
- Remove the danger OR move the casualty.

RESPONSE
- You need to find out if the casualty is conscious or unconscious.
- Look at the casualty and talk to them as you approach them, introduce yourself even if they don't appear to hear you.
- Ask questions about what has happened.
- Give a command - 'Open your eyes'
- If there is no response, gently shake the casualty's shoulders.
- No response – unconscious.

AIRWAY
- Open the airway and if necessary, clear the casualty's airway.
- Place one hand on the casualty's forehead and gently tilt the head back.
- Place two fingertips of your other hand on the point of the casualty's chin and lift the chin.

BREATHING
(check for 10 seconds)
Keep the airway open as you:
Look for chest movement.
Listen for sounds of breathing.
Feel for breaths against your cheek.

THE CASUALTY IS NOT BREATHING
- Ask someone to call an ambulance.
- Give clear instructions.

Note. If you are on your own, make the call before CPR.

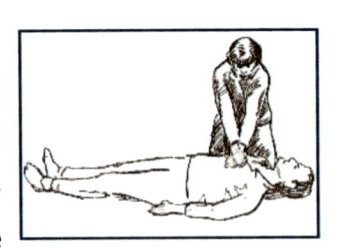

Begin CPR with chest compressions:
- Kneel beside the casualty level with the casualty's chest.
- Place the heel of one hand on the centre of the casualty's chest.
- Place the heel of the other hand on top of the first hand.
- Interlock your fingers, making sure the fingers are kept off the ribs.
- Lean over the casualty with your arms straight.
- Press down vertically on the breastbone and depress the chest by 5-6cms.
- Release the pressure without lifting your hands from the casualty's chest.
- Allow the chest to come back up fully before giving the next compression.

GIVE 30 COMPRESSIONS AT A RATE OF 110 - 120 PER MINUTE.

Please Note. Rescue Breaths (Only undertake this this action if you have been trained, otherwise continue with chest compressions only).

After 30 chest compression, give two rescue breaths as follows:
- Pinch the person's nose.
- Tilt the chin up using two fingers.
- Place your mouth over their mouth and, blowing steadily, attempt two rescue breaths each over one second.
- Keeping head tilted and chin lifted, look down to see the casualty's chest fall.

Repeat chest compressions without delay. Continue the cycle of 30 chest compressions followed by 2 rescue breaths until either:
- Emergency help arrives and takes over.
- The casualty starts to breathe normally.
- You become too exhausted to continue.

THE CASUALTY IS UNCONSCIOUS BUT IS BREATHING
- Don't move the casualty unless there is further danger.
- Check for any life threatening injuries such as severe bleeding and treat.
- Don't try to remove a motor cyclist's helmet unless there are breathing problems.
- Put the casualty in the **recovery position**.

The Recovery Position
- Kneel beside the casualty.
- Remove spectacles or any bulky items from pockets.
- Make sure both legs are straight.
- Place the arm nearest you at right angles to the casualty's body with bent elbow and palm upwards. Bring furthest arm towards you and hold the back of his hand against the cheek nearest to you.
- With your other hand, grasp the far leg just above the knee and pull it up, keeping the foot flat on the ground.
- Keeping the casualty's hand pressed against his cheek, pull on the far leg and roll the casualty towards you.
- Make adjustments to the upper leg so that both the hip and knee are bent at right angles.
- Tilt the casualty's head back and tilt the chin to keep the airway open.
- Call 999 if not already done and monitor the casualty (breathing, response and circulation).

Severe Bleeding – Precautions
- Do not apply a tourniquet.
- If there is an embedded object in the wound, apply pressure on either side of the wound, and pad around it before bandaging.
- Wear gloves if available.
- If the casualty loses consciousness – **A B C**.

Severe bleeding - Action
- Apply pressure to the wound.
- Raise and support the injured part.
- **APPLY PRESSURE AND ELEVATE**
- Bandage the wound.
- Call ambulance.
- Treat for shock and monitor the casualty.

Shock - recognising it
- Rapid pulse.
- Pale, cold, clammy skin.
- Sweating.

Later...
- Grey-blue skin, especially inside lips.
- Weakness and giddiness.
- Nausea or thirst.
- Rapid, shallow breathing.
- Weak pulse.

Shock - Action
- Help casualty to lie down and raise their feet.
- Loosen any tight clothing and cover the casualty with a blanket or coat.
- Call an ambulance or ask another person to do this.
- Monitor the casualty's breathing, and response.

Heart attack – recognising it
- Vice-like pain in the chest spreading to one or both arms.
- Breathlessness.
- Discomfort, like indigestion, in the upper abdomen.
- Sudden faintness.
- Collapse.
- Sense of impending doom.
- Ashen skin and blueness at lips.
- Pulse goes from fast to weak.
- Extreme sweating.

Heart attack – precautions
- Do not give any fluids/drinks.
- If the casualty loses consciousness – ABC.

Heart attack – Action
- Make casualty comfortable.
- Help the casualty into a sitting position.
- Support his head, shoulders and knees (cushions, pillows or bulky clothing).
- Reassure him.
- Call for an ambulance.
- Give casualty an aspirin to chew slowly (only if he is conscious).
- If casualty has his own tablets or puffer aerosol for angina, allow him to administer it himself, but help if necessary.
- Monitor the casualty.

Broken Bones - Recognition
- Swelling, distortion and bruising.
- Pain.
- Difficulty in moving the injured part.
- There may be some bending or twisting of a limb.
- There may be a wound with a bone protruding.

Broken bones - precautions
- Do not try to move an injured limb unless absolutely necessary.
- Do not try to bandage the injury if medical assistance is on the way.
- Do not let the casualty eat, drink or smoke.

Broken Bones - Action
- Help the casualty to support the injured part above and below the injury in the most comfortable position.
- Protect the injury with padding.
- Call an ambulance if the casualty cannot be taken to hospital.
- Treat the casualty for shock.
- Monitor the casualty.

Questions on First Aid
Q1. The first aid kit in your bus should contain...
    A. A book on First Aid
    B. A range of dressings
    C. Splints for broken bones
    D. A tourniquet to arrest any bleeding

Q2. One sign that a bone may be broken is...
    A. The lack of pain in that area
    B. The casualty may ask for a drink of water
    C. The casualty may have some swelling, distortion and bruising in that area
    D. The casualty may have difficulty speaking

Q3. When dealing with a seriously injured casualty it is helpful to remember.
    A. Nurse 1,2,3
    B. Ambulance A, B, C
    C. Dr, A, B, C
    D. Nurse A, B, C

Q4. When approaching a casualty who is unconscious you should...
   A. Start C P R
   B. Speak to them; ask if they can hear you
   C. Help the casualty into a sitting position
   D. Offer them a drink of water

Q5. You have to attend to a casualty who is both unconscious and not breathing, you should...
   A. Ask someone to call an ambulance or if on your own call 999 yourself
   B. Run and get help
   C. Try and find some disposable gloves
   D. Move them to a more comfortable position

Q6. You have to attend to a casualty who clearly has a broken leg, you should...
   A. See if they can put their weight on it
   B. Ask where it hurts
   C. Move them to a more comfortable position
   D. Protect the injury with padding and call for an ambulance

Q7. You arrive at an accident where a casualty has a deep wound and is bleeding, you should...
   A. Get a plaster from the first aid kit
   B. Get someone to telephone for an ambulance and if possible raise and support the injured part
   C. Look away
   D. Tell your passengers to walk home as you are going to be a while

Q8. An elderly person on your vehicle has taken a nasty fall, you should first...
   A. Tell them to get up as you are already late
   B. Help them to get up
   C. If conscious, keep them still and establish if they have any injuries
   D. Telephone the garage and tell them that you will be late

Q9. An elderly person on your vehicle has been taken ill, you suspect a heart attack, you should...
A. Begin CPR with chest compressions
B. Have a look and see what you have in the first aid kit
C. Get someone to telephone for an ambulance, make casualty comfortable and reassure them
D. Get the casualty to lie down

Q10. You have to attend to a casualty who clearly is in shock, you should...
A. Place them in the recovery position
B. Loosen any tight clothing and cover the casualty with a blanket or coat
C. Tell them to pull themselves together
D. Ignore them

MANUAL HANDLING

As drivers of large vehicles (either LGVs or PCVs) it is inevitable that you will be required to lift heavy objects from time to time, if not regularly. As such, it is important that you adopt the correct techniques for doing so. Failure to adopt correct techniques can result in strains or injury which could require time off work to recover. Of those who take more than three days off work due to injury over one third are caused by strain or injury due to incorrect manual handling, resulting in a significant number of days lost from the workplace and pain and suffering for all those who sustain injury.

When Manual Handling, employees should:

1. Adhere to appropriate safe working practices.
2. Make full and correct use of equipment provided for your safety and that of others.
3. Let your employer or line manager know immediately if you identify hazardous handling activities.
4. Try to ensure that your actions in the workplace do not put you or others at risk.

Steps to follow for Safe Manual Handling

- Think first before lifting.
- Will I injure myself if I try to lift the load?
- Where am I going to put the load? Is the route clear?
- Do I need help?
- Is there an easier way of moving this object?
- Are there any tools or machines available to help me do the job?

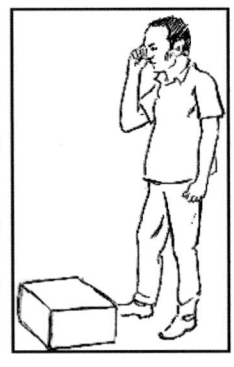

- If I have to lift, I will need to take up a stable position with my feet apart to help me keep my balance throughout the lift.

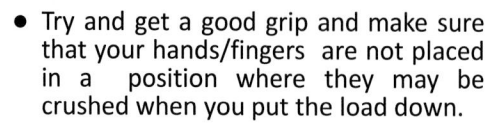

- Try and get a good grip and make sure that your hands/fingers are not placed in a position where they may be crushed when you put the load down.

- Keep the load as close as possible with the heaviest part nearest to your body.

- When starting to lift, if the load is on the ground or low down, try to avoid bending your back, bend at the knees and hips .

- Try not to lean or twist sideways as this can cause injury while the back is bent. If you have to turn, use your feet not your body.

- Keep your head up when lifting and moving. Keep looking ahead, not down at the load.

- Do not attempt to carry a load that obstructs your line of sight.

- If the load can be split into smaller parts, don't lift more than you can handle. Especially if you have had health issues in the past.

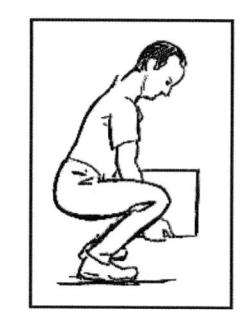

- Don't try and be a hero. If you need help, ask for it. It's a lot less painful than a week in bed with a bad back.

- If you have to move a heavy object, far better to work as a team to get the job done without injury.

- Think! Is there is a tool or piece of equipment available to help you do the job, if so, use it.

- Why place yourself at risk of injury when there may be other solutions available and easier ways to do the job.

- Make it easy on yourself. Before you attempt to lift, ensure that you move any obstructions that you could fall over or that may prevent you lifting in a safe manner.

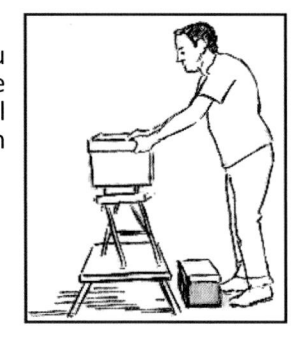

Remember! Think safe - be safe

For information on manual handling you can go to the HSE website.

Questions on Manual Handling

Q1. When considering lifting a heavy object, you should...
   A. Undertake a weight training course
   B. Think and look for the safest way to do it
   C. Just get on with it
   D. Let others struggle

Q2. You have been asked to go on a manual handling course by your employer, you should...
   A. Refuse as you think that it's a load of rubbish
   B. Just consider it a day off
   C. Try and learn as much as possible
   D. Just go for the certificate

Q3. You need to move a large object, do you...
   A. Just get on and lift it
   B. Look for help
   C. Get a younger person to move it
   D. Take a break and hope it's gone when you get back

Q4. When lifting or moving heavy objects employees should consider...
   A. If lifting equipment is available to do the job
   B. If anyone else can do it
   C. If the boss is looking
   D. If the item has a bar code on it

Q5. Drivers should only lift heavy objects when...
   A. No one is looking
   B. The object is covered in protective packaging
   C. Help is at hand or lifting equipment is available
   D. They are feeling well

Q6. Lifting of heavy objects is a problem for...
   A. The bosses
   B. Only the staff
   C. For all, managers and drivers
   D. The people who own the objects

Q7. You see a colleague trying to lift a very heavy object on his own, you should...
   A. Stand back and watch
   B. Give instructions
   C. Go and get a cup of tea
   D. Give assistance

Q8. You are asked to move a heavy box and the sack-truck is on the other side of the building, you should...
A. Go and get it
B. Not bother as it's not worth the walk
C. Try and lift it on your own
D. Wait until someone else moves it

Q9. You become aware of a situation which could result in strain or injury. You should...
A. Do nothing
B. Stop the activity and get help if required
C. Just give advice
D. Tell the boss next time you see him/her

Q10. Before attempting to lift a heavy item, you should...
A. Take a tea break
B. Consider the safety of others who may be in the area
C. Tell people to get out of the way
D. Just get on with it

PASSENGER SAFETY AND COMFORT

In order to ensure that your passengers are safe throughout their journey you will need to have understanding and knowledge of the following:

- Ensuring the safety and comfort of your passengers throughout their journey.
- Understanding the forces affecting your vehicle and its passengers.
- Managing your passengers and conflicts.
- Seatbelt Regulations.
- The carriage of people with disabilities or special needs.
- Understanding the regulations governing the carriage of passengers in the UK.
- Understanding the regulations governing the carriage of passengers internationally.

ENSURING THE SAFETY AND COMFORT OF PASSENGERS
As a professional driver your first priority is to ensure the safety and comfort of your passengers throughout their journey. So what do you need to do to achieve this goal?

The key to all that you do and achieve is attitude. It sounds simple but having a positive attitude to your work and passengers will shine through.

What practical steps can you take to enable your passengers to have a safe, comfortable and trouble free journey?

1. Make sure that you have completed your daily checks and your vehicle is safe to drive.
2. Make sure that you are fit to drive.
3. Plan your journey.
4. Be in control of yourself and your vehicle at all times.
5. Be patient with others and drive with care.
6. Ensure the safety of others.
7. Be observant and try to anticipate events.
8. Load all luggage safely with no overloading of the vehicle or axles.

Daily Checks
Completing your daily check list and ensuring that any faults are attended to is key to ensuring that you have a trouble free journey and that your vehicle is fit to drive. The daily check list is covered elsewhere in this booklet.

Fitness to Drive
You must ensure that before you start your journey you are fit to drive your vehicle; ensure that:
i. You have rested and had enough sleep.
ii. You feel well and have no major ailments that could affect your driving.
iii. You are not on any medication that can affect your ability to drive.
iv. You are relaxed and not stressed.
v. You have not consumed any alcohol or prescribed drugs that could affect your driving.

Planning your Journey
When planning your journey you need to ensure the following:
i. You know where you are going.
ii. You're satisfied that you have the best possible route to get there.
iii. You have checked as far as possible against traffic holdups.
iv. You have planned stops for breaks.
v. If travelling with other vehicles, all drivers are following the same route.
vi. You use dedicated bus lanes in traffic to save time.

Being in control
Being in control of yourself and your vehicle at all times is essential. **You should not...**
i. React to aggression by others.
ii. Be aggressive with or towards other road users or your vehicle.
iii. Be goaded and back away when others are aggressive towards you or your vehicle.
iv. Be competitive with other road users by racing your vehicle.
v. Take risks or show off with your vehicle.
vi. Make sudden movements with your vehicle. These include acceleration, braking and cornering. Passenger safety and comfort is your priority.

1. **Be patient and drive with care**
   Being patient with others (passengers, colleagues and other road users) allows you to drive with care. You need to remember at all times that you are the professional driver. You should...
   i. Be grateful that other people's failings are not yours.
   ii. Allow for their mistakes.
   iii. Be content, you don't have to prove yourself by being competitive.
   iv. Consider others before yourself.
   v. Always drive to the best of your ability.

2. **Ensuring the safety of others**
   Ensuring the safety of others is your first priority. This includes other road users as well as any passengers and colleagues. You should ...
   i. Ensure that your passengers are seated before moving off.
   ii. Where possible discourage your passengers from leaving their seats whilst the vehicle is moving.
   iii. If fitted, encourage the wearing of seatbelts for adults (advise them that it's the law). Those under 14 years on the school run may need a firmer prompt.
   iv. Always drive with care taking account of road conditions.
   v. Always be aware of pedestrians and other road users.

3. **Be observant and try to anticipate events**
   Being observant and anticipating events can prevent you getting into trouble. In order to develop those skills you should...
   i. Always plan ahead for the next road hazard.
   ii. Always remember that as a professional driver many other road users simply don't have your skills.
   iii. Consider the size of your vehicle and the position of more vulnerable road users.
   iv. Anticipate that other drivers may do unexpected things.
   v. Adopt a safety first approach to everything you do.
   vi. Loading your vehicle safely with no overloading of the vehicle or its axles.

4. **It is important that you load your vehicle with care to ensure the following...**
   i. That you don't overload your vehicle by exceeding its maximum permissible weight.
   ii. That you don't exceed the maximum permissible weight for each axle.

iii. If making a number of stops, ensure as far as possible that the luggage is stowed in such a way as to give access to luggage at each stop.
iv. Encourage passengers to sit throughout the vehicle and not all on one side or on the top deck of a double decked vehicle.
v. If driving a service bus, ensure that items of luggage are stowed safely before moving off.
vi. Ensure wheel chair passengers are secure in the vehicle.
vii. If in a service bus don't allow too many standing passengers on your vehicle.

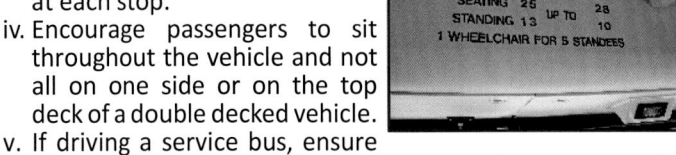

Questions on Passenger Safety and Comfort

Q1. Planning ahead will help you and others stay safe.
    A. True
    B. Not true as it takes your mind off what you are doing
    C. Only if travelling at speed
    D. Only true when travelling on long journeys

Q2. Being competitive with other drivers improves safety.
    A. True as this is one way to improve your driving skills
    B. Only when travelling on motorways
    C. Not true as being competitive increases risk
    D. True as it helps the time go by on long journeys

Q3. You should always assess your fitness to drive before starting out...
    A. No, because if you don't drive you don't get paid
    B. No because if you feel a little unwell it won't affect your ability to drive
    C. Yes because if you have been drinking you may be over the drink drive limit
    D. No because taking any medication won't affect your ability to drive

Q4. Being patient with others, even when under pressure is a good quality to have...
    A. No because others will take advantage of you
    B. Yes as it reduces stress and risk
    C. No as your boss won't appreciate you being late
    D. No because there are so many idiots around

Q5. As a driver you should always take account of road conditions...
   A. Only when driving in winter
   B. Only when driving in fog
   C. Yes as all driving conditions affect the safety of your vehicle
   D. Yes because I can go faster in daylight

Q6. Planning your route before starting off will reduce the risk of traffic holdups.
   A. Only during the night
   B. Only on long journeys
   C. Only when travelling with other vehicles
   D. This is true as it may help you avoid any bottle-necks at peak times

Q7. Being a professional driver requires a higher level of skill.
   A. Not true anyone can drive a large vehicle
   B. Only true if pulling a trailer
   C. True as higher standards are required at all times
   D. Not true, you only have to look at some of the others out on the road

Q8. Undertaking your daily check can reduce the risk of breakdowns...
   A. Only if it's not your regular vehicle
   B. True as the vehicle may have a serious fault
   C. Not true as it's just not that important
   D. True but only once a week

Q9. When other road users are aggressive towards you, you should...
   A. Respond with a gesture or hand signal
   B. Respond with aggression
   C. Not respond at all
   D. Use your mobile phone to inform the police

Q10. Being observant and anticipating events can reduce accidents...
   A. Only in bad weather or at night
   B. Anticipation and being pro-active is always better than having to be reactive to events
   C. Anticipating events will only slow you down
   D. Thinking ahead stops you listening to the radio

FORCES AFFECTING YOUR VEHICLE AND PASSENGERS

As previously stated, the safety and comfort of your passengers is the driver's first priority.

Understanding how your vehicle is going to behave when subjected to the following forces is the first step to ensuring a safe and comfortable journey.

- Acceleration.
- Braking.
- Cornering.

Your passengers won't appreciate being thrown around in their seats or be frightened. In addition, hearing their unrestrained luggage crashing around in the hold when being driven at speed around corners, failing to spot a hazard or having to brake hard at the last moment will not instil confidence in you as a driver.

Drivers must remember that they have an absolute responsibility to ensure that the passengers and load (luggage) they are carrying on their vehicle are safe and secure, even if they have not loaded the vehicle themselves. When loading and unloading the vehicle, drivers should ensure that:

1. The load will not cause a danger to themselves or others.
2. If possible the load is distributed evenly over the whole of the bed of the vehicle's luggage area.
3. That any heavy items are carried at the bottom of the load. (All heavy luggage should be placed in the hold or in floor mounted luggage racks, not overhead).
4. Where possible they have sufficient and suitable restraints to secure the load.
5. If a multi drop load, they will need to take account of: -
    i. The order of the drops.
    ii. Access to each part of the load
    iii. Risk when opening cargo doors.
    iv. Personal risk when lifting luggage.

Tips for driving when loaded
1. Drive smoothly at all times.
2. Avoid heavy braking
3. Avoid any rapid or severe changes in direction.
4. Reduce your speed before tight bends or roundabouts.
5. Avoid any rapid acceleration.
6. Remember you will take longer to stop when loaded.

Load Stability VOSA Report
A report by VOSA (Vehicle and Operator Services Agency) identifies that the overloading of vehicles has become a major problem. The report states that in 2008/9 when VOSA weighed 20,218 vehicles at the roadside, a total of 8770 = 30% were found to be overloaded.

Results from 2005/6 showed that at that time some 23% of vehicles checked were found to be overloaded. It would appear that VOSA enforcement is getting tougher as the numbers being tested is going up.

Year	Number Weighed	Number Overloaded	Percentage
2008/9	29218	8770	30%
2005/6	25271	5945	23%

The HSE believes that the problem does not end there. If you overload your vehicle it will affect the stability of your vehicle. Also it is possible to load your vehicle in such a way that it may not exceed the maximum weight of that vehicle but may exceed the individual axle weight. Even when the vehicle is not overloaded but is loaded badly or if the load is not secured properly, this will affect the stability of that vehicle.

Drivers must be aware that they are responsible in law if they drive a vehicle that is overloaded. The driver can find that in addition to the offence of overloading the whole vehicle, they may have further penalties imposed for overloading one or both of the axles. With the maximum fine for each offence being £5,000,

Energy and forces acting on the vehicle and its load
It was Sir Isaac Newton who identified the laws of motion and how things behave when subjected to forces. When a vehicle and its load are moving, both are subject to considerable amounts of force. Drivers need to understand what these forces are and how they can affect the stability and safe operation of the vehicle.

Gravity
As we all know gravity is a force that acts downwards on the vehicle and anything it may be carrying. The heavier the vehicle the more it is affected by this gravitational force and more energy is then required to move from rest and to stop it.

Friction
Without this force, the vehicle could not move. (Friction is needed between the tyre and the road and clutch plate for the vehicle to move.) Also, without friction, the vehicle would not be able to stop. (Friction is needed between the tyre and the road and the friction between the brake shoes and brake drum).

Acceleration
In order to make the vehicle move energy is required. The faster it moves or the quicker it accelerates the more energy is required. This energy usually comes from the engine and the fuel it burns. Once the vehicle is moving it has Kinetic energy or if at the top of a hill it has Potential energy.

Braking
When braking, particularly if the vehicle is heavy and the braking is fierce, passengers can be thrown about. Those standing on service buses can lose their balance, fall over and suffer possible injury. Those passengers not wearing seatbelts can be thrown out of their seats.

Cornering
When a vehicle is cornering, a different type of force is acting on the vehicle and its passengers. This type of force is called Centrifugal force and this wants to make the vehicle move out away from the bend. Again, if travelling too fast the vehicle can become unstable.

Travelling
When travelling uphill the driver will find that he/she has to drive the vehicle harder to overcome the gravitational force acting on the vehicle and its passengers.

However, less braking will be needed to stop. Drivers will also know that when travelling downhill the opposite is true. The driver will need to brake harder but use less power to make the vehicle move.

Wind

Driving in wind is a big problem for drivers, particularly if you have a high sided vehicle (Double or Semi-Decker). It is not uncommon for some roads and bridges to be closed or have speed restrictions placed on them during periods of high winds. Each year we see vehicles toppled over due to high winds. Empty vehicles on exposed or open roads tend to be affected more by wind.

Questions on understanding the forces

Q1. When driving your vehicle, who has responsibility for the safety of the passengers?
  A. The boss
  B. The passengers
  C. The person who hired the coach
  D. You the driver

Q2. You believe the load (luggage) is starting to move about part way through your journey. You should...
  A. Check it when you next stop for a break
  B. Pull over at the next safe place and check
  C. Wait until you reach your destination
  D. Ring the boss and ask what to do

Q3. When loading a mixed load with heavy items (cases) you should place...
  A. The heavy ones on top to stop lighter items moving around
  B. The heavy ones at the back so it is easier to get them on and off
  C. The heavy ones on the bottom so that they're less likely to move
  D. Any way you like as long as you get them in

Q4. Constant and rapid changes of direction can...
  A. Be fun for the passengers
  B. Cause travel sickness
  C. Improve fuel consumption
  D. Improve the driver's image

Q5. Who has responsibility for the safe loading of luggage on to your vehicle?
   A. The passengers
   B. The driver
   C. However puts the luggage on the bus
   D. The manager of the company

Q6. Continuous rapid acceleration of your vehicle can result in
   A. More profit for the company
   B. Using far more fuel for the journey
   C. More passengers getting to their destination quickly and safely
   D. Getting yourself a bonus

Q7. Single deck coaches are unaffected by high cross winds...
   A. This is true
   B. Only at night
   C. High crosswinds can be dangerous even for single deck coaches
   D. High crosswind will only affect the coach when empty

Q8. When driving in poor weather conditions such as ice and snow, friction between the wheels and the road improves....
   A. This is true
   B. Only when the vehicle is empty
   C. Not true, friction between the tyre and the road is reduced
   D. Only when driving fast

Q9. To ensure that passengers have a safe and smooth journey, drivers should...
   A. Take account of the weather conditions at all times
   B. Drive as fast as their vehicle can go at all time regardless of the conditions
   C. Drive slowly at all times
   D. Always drive with the wind behind them

Q10. Overloading your vehicle could result in...
   A. A fine for the driver
   B. A fine for the company
   C. Better cornering throughout the journey
   D. A fine for the driver and the company

SEAT BELT REGULATION

In general the law states that if a seatbelt is fitted it must be used. However, in practice it's a bit more complicated than that and all drivers, particularly coach and minibus drivers, need to be aware of the regulations.

So what are the regulations on the wearing of seatbelts?

    1983 The wearing of seatbelts in cars became compulsory

    2006 Further regulation was introduced covering the wearing of seatbelts in cars, small minibuses under 2540 kg, light vans, medium and heavy vehicles.

\multicolumn{4}{c}{Occupants of cars vans and goods vehicles}
Vehicle occupant	Front seat occupant	Rear seat occupant	Who is responsible?
Driver	Seatbelt must be worn if fitted		Driver
Adult passengers 14 years and older	Seatbelt must be worn if fitted	Seatbelt must be worn if fitted.	Passenger
Children up to 1.35m (approx 4 ft 5 in) or 12 to 13 years old	The correct child restraint must be used.	If a seatbelt is fitted the correct child restraint must be used.	Driver
Children up to 3 years old	The correct child restraint must be used	The correct child restraint must be used. However, if one is not available (taxis only) then they can travel unrestrained. This regulation does not apply to any other vehicle.	Driver

Vehicle occupant	Front seat occupant	Rear seat occupant	Who is responsible?
Children more than 1.35 m (approx 4 ft 5") or 12 years of age until 14th birthday.	The seatbelt must be used.	Where there is a seatbelt fitted it must be used	Driver

Occupants of buses and coaches including mini buses over 2540kg

Vehicle Occupant	Front Seat	Rear Seat Occupant	Who is responsible?
Seated passengers 14 and older	Seatbelt must be worn if fitted	Seatbelt must be worn if fitted	The passenger
Children aged 3 to 13 years of age	The correct child restraint must be used	There is currently no legal requirement for passengers to wear seatbelts	Driver

Child Restraints
Modern restraints are designed to fit the weight of the child and have to meet UN and EU regulations. Rear facing baby seats should not be used where there is an air bag. Whatever the regulations, the wearing of seatbelts saves lives. The statistics show that after the compulsory wearing of front seatbelts came into force the number of deaths and casualties from vehicle occupants dropped significantly, this was followed by a further drop after the introduction of the compulsory wearing of rear seatbelts.

Exemption Certificates
It is possible to obtain an exemption certificate from your doctor allowing you to travel in a motor vehicle without having to wear a seatbelt. In such cases the certificate must be carried with you each time you travel.

Penalty for not wearing a seatbelt
The current penalty for not complying with the regulations is a maximum fine of £500. However, you will not be awarded points on your licence and it is more usual to receive a fixed penalty notice of £60. That said the cost of not wearing a seatbelt if involved in an accident could be significantly greater.

Questions on Seat Belt Regulations

Q1. Who is responsible for ensuring that an adult seated in a rear passenger seat wears his seatbelt?
A. The driver
B. The passenger
C. It is not a legal requirement
D. The police

Q2. The police stop you for not wearing a seatbelt when driving a coach fitted with a seatbelt. You are…
A. Not committing an offence
B. Only committing an offence if driving on a motorway
C. Committing an offence
D. Committing an offence which will place points on your licence

Q3. Adults need not wear seatbelts if fitted in a coach.
A. This is true
B. Only if they are travelling on short journeys
C. They are required to wear them by law
D. Only required by law for front seat passengers

Q4. Who is responsible for ensuring that adults wear their seatbelts when travelling in your vehicle?
A. You the driver
B. The owner of the vehicle
C. The police
D. The passenger

Q5. It is legal for children less than 3 years of age to be carried in the front of vehicles.
A. Only if the vehicle is a taxi
B. Only if the correct child seat is fitted
C. Only if sat on an adult's lap
D. Only if the driver is the child's parent

Q6. Who is responsible for ensuring that children under 14 years wear their seatbelt when travelling in your car?
A. The parent of the child
B. The child
C. The police
D. The driver

Q7. An adult may travel unrestrained in the rear of a van when no seats or seatbelts are fitted...
A. This is true
B. This is true but only for short journeys
C. It is illegal as there is no seat fitted
D. Only if the driver agrees

Q8. Whilst travelling in the rear of a coach, it is not illegal for children aged between 3 and 13 years of age not to wear their seatbelts even if fitted.
A. This is true but unadvisable
B. Untrue and they must wear their seatbelts by law
C. It depends on the size of the coach
D. Only if the coach has a single deck

Q9. There is no evidence to show that seatbelts save lives.
A. This is true but people think that they are safer
B. Seat belts can cause injury and should not be worn
C. There have been many studies done which prove the case for seatbelts
D. The evidence shows that seat belts only work for front seat occupants

Q10. Drivers should not place a rear facing child seat in a seat that is fitted with a front activated air bag.
A. Not true as the air bag makes it safer for the child
B. Makes no difference
C. It is illegal for children to be carried in the front seat
D. It is dangerous because if the air bag is deployed it can injure the child

THE CARRIAGE OF PEOPLE WITH DISABILITIES OR SPECIAL NEEDS

In 2000 new regulations came into force so that all new buses and coaches had to be designed to allow for the carriage of people with special needs. These include people with disabilities, small children or infants and the elderly. Figures show that during our lifetime, over 25% of us will have some sort of disability at some time or other. So to discriminate or exclude disabled persons from using your vehicle is to turn away 25% of your customers.

In 2006 further legislation came into force which gave people with disabilities the right to be able to access transport services.

This legislation has made it illegal to discriminate against people with disabilities. That said it's not about the law. It's about doing the right thing.

What to do
- Firstly treat people with respect.
- Give those with disabilities as much time as possible to get themselves organised.
- Give them time to reach a seat, a simple thing but not doing so can create anxiety for them.
- If required assist those who may need help or support.
- Use your senses. Not all disabilities are immediately obvious. Deafness is an example.
- If your vehicle does not have the ability to kneel, try and stop your vehicle as close to the kerb as possible as they may not be able to cope with high steps on to your vehicle.
- If your vehicle has a lift to allow people in wheelchairs to board whilst still seated, you should...
  1. Have previously undergone training (Midas) or similar.
  2. Check the lift can take the weight of the chair.
  3. Give assistance.

    4. Ensure that the wheelchair is placed in the correct position and secured correctly.
    5. Talk to the occupant to give them reassurance.
- Some people may not be classed as disabled but may need some assistance. Examples of this may be a mum with small children and a push chair, or older people.
- Be patient, it might be you one day!

What not to do
- Do not patronise or demean those who may have less ability than you.
- Don't discriminate.
- Don't be impatient.
- Many people may suffer from panic attacks, not just those with a disability. Therefore don't take risks and drive with extra care so as not to frighten them.
- Some disabilities may also cause discomfort which may restrict the ability to move easily. So don't drive off until the passenger has sat down or drive in a way that causes any rapid changes in direction, acceleration or braking.
- Don't try to help them if they say that they are fine. We all value our independence so let people help themselves if they wish to, unless it's unsafe.

If you are carrying people who use wheelchairs, you will need special training in order to support them, operate equipment, secure their chairs in place and enable them to get on and off your vehicle.

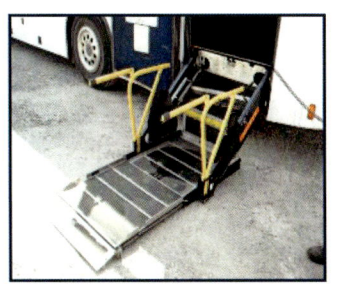

Questions on the carriage of people with disabilities and special needs

Q1. When a young mum with a pushchair and three children attempts to board your bus you should...
    A. Do nothing
    B. Stop her and the children from boarding the vehicle
    C. Lend a helping hand and stow the pushchair safely
    D. Tell her to get to her seat as quick as she can

Q2. You have an elderly nervous passenger on your bus. You should...
   A. Drive faster
   B. Tell her she has to get off at the next stop
   C. Tell her to stop complaining
   D. Drive with a little more care and reassure her that all is fine

Q3. A person with learning difficulties has difficulty understanding money. Should you...
   A. Tell them to get off the bus
   B. Give them time and assistance
   C. Drive on without taking the money
   D. Charge them double

Q4. An elderly man who seems a little unsteady on his feet is taking some time to board your vehicle. You ask if you can help but he says no. You should...
   A. Ask him to get off the bus
   B. Help him anyway
   C. Just give him a little more time and space
   D. Tell him to get a move on as you are already late

Q5. You see a person in a wheelchair waiting at the bus stop. You are aware that the person can board your vehicle without the chair. You should...
   A. Stop, help them and the chair on to the bus and place the chair in a safe place
   B. Drive past and pretend you did not see them because you are late
   C. Phone the driver of the next bus and tell them to pick them up
   D. Stop and tell them they can't get on

Q6. Whilst driving your bus with a number of able bodied and disabled people on board, you notice smoke coming from the engine compartment. You should...
   A. Drive to the nearest fire station
   B. Stop the bus and jump out as fast as you can
   C. Stop the bus and get everybody off, giving help where required
   D. Stop the bus and phone 999 for the fire service

Q7. A confused young child has boarded your bus but does not know where he/she needs to get off. Do you...
   A. Just put them off at the next stop
   B. Keep them with you until you reach the terminal
   C. Hand them over to a stranger on the bus
   D. Phone the police for help

Q8. An elderly and infirm person has just got on to your vehicle and is making their way to a seat. Do you...
   A. Wait for them to take their seat
   B. Drive off as quickly as you can
   C. Drive off slowly
   D. Shout and tell them to hurry up

Q9. A passenger is taken seriously ill on your bus. Do you
   A. Put them off at the next stop
   B. Ring for an ambulance
   C. Jump a red light to get them to a hospital
   D. Ignore them and hope they get better

Q10. On approaching a bus stop you see a passenger who has passed out on the pavement. Do you...
   A. Drive past
   B. Stop and tell them to get up or you'll be late
   C. Stop, give help and phone for an ambulance
   D. Ring the depot and tell them you are going to be late

HOW TO ASSESS AND DEAL WITH EMERGENCY SITUATIONS

Emergency situations can happen at any time, so what could be regarded as an emergency?

Let's look at some incidents that might be regarded as emergency situations.
   A. A road traffic accident involving injury.
   B. Someone becoming seriously unwell.
   C. A breakdown in a dangerous situation.
   D. A fire on your vehicle.

A. A road traffic accident involving injury. What should you do?

1. Firstly stay calm. If you panic so will others.
2. Take control. As the professional you are in charge until the emergency services arrive. If you have a fluorescent tabard, put it on so people can see you.
3. If your vehicle is not involved, stop your vehicle in a safe place. If you are first to arrive at an accident scene, warn other drivers and contact the emergency services before giving help. Only if it is safe to do so, try to protect the scene with your vehicle.
4. If your vehicle is not involved, it may be safer not to evacuate your passengers.
5. If your vehicle is involved, first check for injuries and if you need to evacuate your vehicle, ensure that you can account for everybody (no one missing) and they are all in a safe place.
6. If possible switch off the fuel supply on your vehicle.
7. Don't allow smoking in case of fuel leaks.
8. Alert the emergency services giving exact details of location and the problem.
9. Do not stand in a position that blocks the view of the rear lights of your vehicle.
10. Check to see if there are any medical professionals on board who may help with casualties. If you are trained, give basic first aid.
11. Reassure and keep reassuring all those you are supporting.
12. If you can, delegate some of the tasks (such as contacting the emergency services) to a responsible person.

B. Someone has become seriously unwell on your vehicle. What should you do?

1. Again stay calm. If you panic so will others.
2. Stop your vehicle in a safe place.
3. Take control. As the professional you are in charge until the emergency services arrive.
4. Alert the emergency services giving exact details of location and the problem.
5. Check to see if there are any medical professionals or first aiders on board who may help the casualty. If trained, give basic first aid.
6. Reassure and keep reassuring all those you are supporting.
7. If you can, delegate some of the tasks (contacting the emergency services) to a responsible person.

C. A breakdown in a dangerous situation. What should you do?

1. Again stay calm. If you panic, so will others.
2. Take control. As the professional, you are in charge until the emergency services arrive. If you have a fluorescent tabard, put it on so people can see you.
3. If safe to do so, try and warn other road users. Put your hazard lights on.
4. Only if safe to do so, evacuate your passengers and yourself to a safe place. Give help to those with special needs or injuries. Ensure that all passengers leave the vehicle via the near side.
5. Don't allow smoking in case of fuel leaks.
6. Alert the emergency services giving exact details of location and the problem.
7. Do not stand in a position that blocks the view of the rear lights of your vehicle.
8. Reassure and keep reassuring passengers.

D. A fire on your vehicle. What should you do?

1. Again stay calm. If you panic so will others.
2. If possible stop your vehicle in a safe place. (However, you may not have time to look for a safe place to stop).
3. Evacuate your vehicle. (This is a priority.) Help those who may need help in getting off the vehicle.
4. Get your passengers to a safe place.

5. Ensure that you can account for everyone (no-one missing). Take control. As the professional, you are in charge until the emergency services arrive.
6. Alert the emergency services giving exact details of location and the problem.
7. If possible switch off fuel and electricity supplies on your vehicle.
8. Do not try to put out the fire if it places you in any danger.
9. If any persons are injured and if trained, give basic first aid. Also, check to see if there are any medical professionals on board who may help any casualties.
10. Reassure and keep reassuring all those you are supporting.
11. If you can, delegate some of the tasks (such as contacting the emergency services) to a responsible person.

Questions on dealing with emergency situations

Q1. You arrive first on the scene of an accident. Do you...
   A. Do nothing
   B. Telephone for help
   C. Jump out and start giving first aid
   D. Drive past if possible

Q2. A passenger on your vehicle is taken seriously ill. Do you...
   A. Drop them off at the next available place to stop
   B. Phone a friend for advice
   C. Contact the emergency services and give first aid if trained
   D. Drive as fast as you can to the hospital

Q3. If a fire breaks out on your vehicle, do you...
   A. Drive to the nearest lay-by
   B. Stop the vehicle as soon as possible and evacuate any passengers and yourself to a safe place
   C. Wind down the windows to get some fresh air
   D. Turn the heater off

Q4. Your vehicle has broken down in a place that could cause a danger to others. Do you...
   A. Stay in your vehicle
   B. Get out and run away
   C. Try and fix the problem
   D. Phone for help and warn other drivers whilst not endangering yourself

Q5. Another driver has driven at speed into the back of your vehicle. Do you...
   A. Just drive on and hope the boss does not notice the damage
   B. Get out and call him an idiot
   C. Get out of your vehicle and assess the scene
   D. Ring your insurance company

Q6. You witness a very nasty accident. Do you...
   A. Drive on so as not to be late
   B. Stop and contact the emergency services
   C. Leave your name and continue on your journey
   D. Telephone your boss first and tell him you are going to be late

Q7. Your vehicle has broken down on the motorway hard shoulder and is in danger of catching fire. Do you...
   A. Leave your passengers on the vehicle and get help
   B. Evacuate them via the near side of the vehicle to a safe place before telephoning for help
   C. Do nothing and wait for the police
   D. Get out and fix the problem

Q8. You have been involved in a non-injury accident. Do you...
   A. Drive off and say nothing
   B. Look around and see if anyone saw you
   C. Give your name and your vehicle owner's name and address and insurance company name (if known)
   D. Dial 999 and ask for the police

Q9. One of your passengers is taken ill and is panicking. Do you...
   A. Slap them on the face and tell them to grow up
   B. Ignore them and hope they stop
   C. Just tell them to shut up
   D. Reassure them and try and help them to calm down

Q10. Whilst on the motorway, you see smoke coming out from under the engine cover. Do you...
   A. Drive faster to try and get to the services
   B. Wind down the window to let more air in to the vehicle
   C. Pull over to the hard shoulder and evacuate the vehicle
   D. Telephone your boss and ask what to do

PROPOSED EVACUATION PROCEDURE (BUSES, COACHES & MINI-BUSES)

Apart from America or the Aircraft industry, research has shown that at the time of writing there appears to be nothing written in law that requires bus operators or their drivers to practice or follow a set procedure in order to evacuate their vehicles in an emergency. However, bus operators and their drivers do have a 'Duty of Care' with regard to all those travelling on their vehicles. As such, they must do all they can to ensure the safety of their passengers.

In all cases the driver will need to think clearly and make judgements regarding the safety of his or her passengers. Every situation will be different.

Drivers need to ask themselves 'would my passengers be safer on or off the vehicle?' and **'Assess the Risk'**.

An example of this would be if you were to have a minor incident or breakdown on the motorway and your bus was full of small school children. Keeping them on the bus may well be the safest option. Small or disabled passengers will not be able to climb over the safety barriers unaided, adding to the risk of them spilling into the carriageway. However, it would be wise to move any passengers out of the back rows of the bus. This is because as many as 30% of all motorway accidents occur on the hard shoulder.

In an incident such as a fire or the vehicle filling with smoke, the obvious and safest option is to get everybody off, well away from the vehicle and if possible, away from the road. However, before undertaking this task the driver must quickly make a full assessment of the situation so as not to expose passengers to greater danger when exiting the vehicle.

Evacuation Procedure
In order to carry out this procedure as quickly and safely as possible, you as the professional driver must be firm and take full control. Before starting your journey you should if possible, try to identify a responsible person on board (This may be a courier, bus prefect, teacher or a member of the public).

It is important to get their agreement that, in an emergency, should the bus need to be evacuated, they will take temporary control of the passengers when they first leave the vehicle. Drivers will need to advise that person of the drill and what will be required of them.

That said it may be that the driver cannot find another person to support and has to take full responsibility for the evacuation on their own. If this is the case, they must give clear instruction to passengers that, after leaving the bus, passengers must move away from the vehicle to a place of safety.

As a driver, what should you do?
1. Stay calm, do not panic, you are the professional in charge.
2. Try and stop your vehicle quickly and as safely as possible in a safe place.
3. Advise the responsible person on the bus that there is an emergency, and get them to take control of the passengers when they leave the vehicle. Try and get that person off the bus first. This is in order to get everyone to a place of safety away from the vehicle (35 metres/100 ft minimum) also if possible away from the road.
4. Open all exits on the near side of the bus away from any traffic. Any exits at the rear or off-side of the vehicle should be a last resort option particularly on a motorway.
5. Give the order to evacuate the vehicle and clear instructions on how this should be done. (Use the P.A. system if available. If not, raise your voice so that everyone can hear what you say). Keep repeating the instructions as some passengers may not have been listening. Try to reassure your passengers throughout this procedure.
6. Advise all passengers to leave any bags and to leave the vehicle quickly, in an orderly way, with those nearest the exits leaving first. Advise all passengers that they need to listen to the instructions of the responsible person after leaving the bus, they must move away from the vehicle (towards the front, away from the rear of the vehicle) and if possible away from the road.
7. Try to ensure that passengers do not panic and push others, in order to get off.
8. Stay on the bus to ensure that everyone gets off as quickly as possible in a safe and orderly way.

9. Give support to any passengers who may have difficulty. If you can, get other able bodied passengers to help where necessary. Drivers need to identify vulnerable passengers as they get on the vehicle.
10. Ensure that everybody is off the bus, before leaving the vehicle yourself.
11. Telephone the emergency services. Take back control of the assembled passengers until the emergency services arrive. If a boarding list is available this should be checked ASAP to ensure that everybody is off, safe and if they have any injuries.
12. If passengers have injuries or are in shock keep them warm, give support and if trained, first aid.

Passengers in wheelchairs

The issue of having passengers in wheelchairs raises additional problems. They may have accessed the vehicle via a lift, with both the chair and the person strapped in. Added to this, they may not have the capacity to free themselves.

If time allows, you may be able to evacuate the passengers as you would if there was not an emergency. However, if the vehicle is on fire or is filling with smoke and is endangering the lives of the passengers, you have to get them off as quickly as possible. In addition, if it's an electrical fire, the lift may not work and it would be unlikely that you would have time to operate the lift manually.

In order to evacuate the vehicle quickly, it may be necessary to extract them from their chairs as there simply may not be time to operate the lift and as stated it may not work. You will need to do this with as much care as possible in order to avoid injury, distress or pain to the person occupying the chair. Talk to them and reassure them throughout, whilst taking this last resort action.

As with able bodied passengers, on exiting the vehicle, disabled passengers must be moved away from the vehicle to a place of safety.

Footnote
In all cases the first priority of the driver is the safety of his or her passengers and not to fight the fire unless by doing so it would aid passengers to escape.

Questions on the Evacuation of Buses, Coaches and Minibuses

Q1. In the event of their vehicle breaking down, before evacuating their vehicle, drivers will need to consider ...
If they will be able to get passengers all back on later
If it will make them late for their next pickup
If evacuation will place their passengers in greater danger
How the boss will react

Q2. UK legislation states that passengers must be evacuated in...
   A. 5 minutes
   B. 2 minutes
   C. There are no regulations on passenger evacuation from buses or coaches
   D. 1 minute

Q3. When evacuating your vehicle in an emergency, drivers should...
   A. Ensure that women and children leave the vehicle first
   B. Ensure that everybody is off before they leave the vehicle
   C. First drive to the nearest police station
   D. First telephone the emergency services

Q4. When evacuating wheelchair passengers in the event of a fire, drivers should...
   A. Try and keep the passenger calm and lift or help them from the vehicle as quickly as possible
   B. Drive to the nearest fire station
   C. First ring the emergency services
   D. Run for help

Q5. Whose responsibility is it to ensure the safe evacuation of passengers from your vehicle?
   A. The passengers
   B. The company
   C. The emergency services
   D. The driver

Q6. Passengers should, where possible, be evacuated through …
   A. The exits on the off side of the vehicle
   B. The exits on the rear of the vehicle
   C. The exits on the near side of the vehicle
   D. The exits on the front of the vehicle

Q7. If your vehicle is filling with smoke or on fire, you should first…
   A. Ring the emergency services
   B. Ring the company and ask what to do
   C. Evacuate the vehicle as quickly as possible
   D. Jump off and run away

Q8. Transport operators and their drivers have a duty of care towards their passengers.
   A. Not true
   B. Passengers need to look after themselves
   C. Only if they have paid
   D. True, operators and drivers have a duty to ensure that their passengers are safe

Q9. In an emergency, drivers need to stay calm and take control in order to try and keep everybody safe …
   A. Until the emergency services arrive
   B. Only until everybody is off the bus
   C. Until the boss gets there
   D. Only until they get off the bus

Q10. When evacuating their vehicle in an emergency, drivers need to …
   A. Be the first off the vehicle to run for help
   B. Shout 'follow me' as they jump off the bus
   C. Call 999 before doing anything else
   D. Take full control and get everyone off safely before themselves

MANAGING YOUR PASSENGERS AND CONFLICTS

Many of you will have experienced firsthand; rowdy, aggressive or even violent behaviour from passengers, especially on a late night service bus returning from a town or city centre.

Having to deal with passengers who have had too much to drink or simply out to cause trouble is unfortunately now a regular occurrence. So much so that many companies have withdrawn their late night services.

Dealing with aggression and conflict is not easy and again a stressful situation. Unfortunately some jobs get more than their fair share of angry people. Customer complaints departments, call centres and bus drivers are amongst the top ones.

It is worth noting here that you are responsible for your vehicle and the safety of all your passengers. You don't want to take any action that may endanger others or yourself.

So how do you deal with a customer who is intent on behaving badly, causing you and other passengers a great deal of stress and anxiety?

1. Don't look for conflict. If someone has spoken to you in an aggressive tone or is simply being difficult, don't respond in a like manner. If you do, it can escalate an already tricky situation. Try to behave professionally at all times.

2. If someone is being aggressive and is out looking for trouble, stay calm, control your breathing.

3. Adopt a non-aggressive posture. Don't invade their space and keep your hands low with your palms open (no fists). Speak softly. Don't raise your voice. Shouting won't help calm the situation.

4. Avoid physical contact. The last thing you should do is to grab or hit the person who is causing problems. If they make a complaint or are injured, you may find yourself being charged with assault or losing your job. Also you don't know if they are carrying a weapon. By grabbing them you may give them a reason to use it.

5. Ask how you can help with any problem they may have. Listen to what they have to say. Keep eye contact. Try to placate them. Be positive, not negative with your responses. Don't get excited. Control your breathing. Keep calm at all times.

6. If you can, use humour. If they think something is funny they won't be aggressive.

7. If all else fails, back away.

If someone is or has been violent towards you or any passenger on your vehicle, use the 999 service and call the police and any other emergency services.

The School Run
The school run is probably one of the most stressful parts of the day that bus drivers have to deal with. The first thing to understand is that you are not alone.

Almost every driver has experienced problems with school runs. Some have experienced major problems which have involved the operator, the local authority and the police. This does not excuse bad or even criminal behaviour from what is usually a small minority of pupils. Let's be clear about this from the start, such behaviour is unacceptable. It's not just the driver's problem – it's everybody's problem, the school, the operator, the parents and the pupils, so you are not alone when faced with such problems.

So what can you as the driver do?
When driving on the school run try and take control of your environment. It's not easy and sometimes almost impossible, if you have 50 or 60 screaming kids on board. But there are things you can do:
- If there are no teachers or support staff on the bus, get your employer to work with the your Transport Officer and the school to get bus prefects appointed. Thereafter, work with them so they understand their role. Give them and the other pupils lots of positive feedback on how things are going.

- Try and get your employers to write to School Transport at County Hall (not just when things are going wrong but when things are going well). Be really positive. Tell them when the pupils on your bus are well behaved and what positive role models they are for the school.

- Smile, be welcoming, adopt good body language and tone of voice. Most will respond positively. Just think, if someone you were dealing with was openly hostile towards you, how would you react? You need to break the cycle of negativity and conflict.

- Be confident about your role. You are not only the adult, but also the professional in charge and the safety of your vehicle and everyone on it is your responsibility. The pupils need to know and understand this.

- Try and show pupils some respect, particularly if they have been helpful or you have had a good journey. You don't have to be their friend, just show a little humanity.

- Think of ways you can reward them. Letting them listen to the radio or CD is just one. Having music playing can also alleviate stress.

What should you do if there bad behaviour?

- If the pupils play up in a way that distracts you from driving, you should stop the bus in a safe place, switch off the engine and explain clearly that their behaviour is unacceptable and must stop. You can also advise that the bus may have CCTV and that any poor behaviour will be seen. If you have bus prefects on board, get them to speak to their peers.

- If pupils continue to disrupt, advise that you will not continue the journey until they all behave. They will be aware that the engine has stopped and that they will be late. Advise that the next action you take will be to contact School Transport and the operator via your mobile, to ask for assistance.

- If the behaviour is such that you believe that it is extreme and dangerous tell them that the police will be summoned and ask your operator or transport office to call the police for help or make the call yourself.

What you must not do.
- Don't scream and shout at them, it doesn't work. Some of the pupils have been shouted at all their lives. Some will try to wind you up just to get a reaction. You don't want to find yourself on YouTube.

- Don't ask them to leave the bus and walk home. If they decide to leave, advise them clearly they should not but do not try to physically stop them if they do.

- Do not invade their space, this can be seen as threatening.

- Do not touch them. Avoid any physical contact. You could be arrested for assault.

Report all incidents to the transport office on your return to the depot.

Somerset County Council have given the following advice to all operators and drivers carrying pupils to and from Somerset's schools.

1. Make every effort to defuse situations and avoid confrontation.

2. The safety of the passengers and other members of the public must be given the highest priority.

3. Hands off! Drivers must not touch schoolchildren in any circumstances other than genuine self-defence, medical need or the prevention of a serious offence or threat to safety.

4. Schoolchildren may not be told to get off the bus or refused entry.

5. No racist or other offensive or abusive language will be tolerated.

6. No threats should be made but warnings can be given. Do not issue warnings unless you intend to act upon them.

7. Do not react to bell-ringing or mild verbal abuse. Don't tolerate swearing or physical abuse.

8. If the bus is in motion and a warning light/buzzer indicates that an emergency exit has been opened, you should immediately stop. Do not drive with the buzzer/light sounding/showing.

9. If there is a risk of a passenger causing damage to the vehicle or endangering themselves or other passengers, you should bring the bus to a halt leaving the doors open and ask them to calm down. If no response say the following:

I have stopped the bus because you are causing a disturbance which makes it unsafe for this journey to continue. I will remain stationary until you have all sat down in your seats and become quiet. I am about to notify my central control who will notify your school and the Passenger Transport unit.

If you do not return to your seats and remain quiet it may be necessary for me to call the police. In the meantime I would advise you to remain on the bus because we have not reached our destination.

The doors of the bus are open but I would strongly advise you not to leave for your own safety.

If still disruptive - call the police for assistance.

Somerset County Council have also produce the following code of conduct for Drivers and Passenger Assistants.

- Ensure that you wear your Identification Badge at all times when working.

- Remember you are representing yourself, your company and Somerset County Council.

- Maintain a courteous and professional relationship, be polite at all times.

- Treat all passengers equally and fairly.

- Avoid all unnecessary physical contact.

- Maintain confidentiality at all times, unless information received may be detrimental to the clients heath or well being, only report your concerns to the Transport Officer.

- Report any concerns about passenger behaviour towards you to your Transport Officer as soon as possible.

- Remember that your actions, no matter how well-intentioned can be misinterpreted.

- Be aware of how your driving may be perceived by others.

Drivers may also find that they have to deal with conflict from other drivers.

Road Rage

The UK has one of the worst records for road rage incidents in the world, with men being more likely to commit an aggressive act than women. Some people feel that when sitting in their cars they can behave badly because they have anonymity i.e. nobody knows who they are.

This is not the same for professional drivers with their large sign written vehicles. The fact is that motoring is about dealing with other people and most importantly yourself. If you lose your own self control you have lost the argument. Drivers need to be self aware. Dealing with and reacting to bad driving is another form of conflict that can be a major cause of stress.

Questions on Managing your Passengers and Conflicts

Q1. When dealing with an aggressive passenger you should always...
   A. Be aggressive towards them
   B. Phone the police
   C. Keep calm and lower your voice
   D. Shout at them

Q2. When a passenger who has been drinking starts a row with another passenger on your bus you should...
   A. Ignore it and just drive on
   B. Pull over and try to calm things down
   C. Drive to the nearest police station
   D. Simply grab the drunk and throw him off the bus

Q3. When dealing with an aggressive passenger your body language plays an important part in resolving the conflict. You should...
   A. Get the first punch in
   B. Ask another passenger to hold him down while you phone the police
   C. Keep calm and try to adopt a non-aggressive posture
   D. Raise your hands to protect your face

Q4. You are not having a good day and a pupil on a school run tells you where to go. You should...
   A. Show him the error of his ways
   B. Swear back at him
   C. Keep calm and simply ask him what has upset him
   D. Get another passenger to throw him off the bus

Q5. A pupil is behaving badly on the school run.
   A. You should ask the biggest boy on the bus to sort him out
   B. Tell his mum and dad
   C. Grab them by the throat
   D. Stop the bus in a safe place and tell him you will not move until he behaves

Q6. When dealing with an aggressive and threatening person you should...
   A. Tell them you are a judo champion and will sort them out
   B. Run away as fast as you can
   C. Keep calm and try to calm them
   D. Get another passenger to throw them off the bus

Q7. Invading another person's space when they are angry, can...
   A. Be seen as an act of aggression towards that person
   B. Be appreciated by that person
   C. Be seen as a way of calming things down
   D. Be seen as the right thing to do

Q8. Another driver blasts on their horn and makes aggressive hand gestures at you from their vehicle.
   A. You should force them off the road as you have the larger vehicle
   B. Force them to stop, get out of your vehicle and explain to them the error of their ways
   C. Ignore them and let them go on their way
   D. Make the same or a similar hand gesture back

Q9. Looking for conflict is...
   A. One way of relieving your own stress
   B. A sure way of finding it
   C. A way of making new friends
   D. Driving up your popularity with your passengers

Q10. Being mentally strong and in control of your own emotions when under stress at work can...
   A. Be seen as a fault in your personality
   B. Be seen as a positive trait
   C. Require treatment from your doctor
   D. Make others think of you as weak

CRIMINALITY AND ITS EFFECTS

Criminality can affect transport operations in one of two ways:
1. Where the transport operator and the driver become victims of crime.
2. Where the driver or operator commits the crime.

Victims of Crime
Figures from the Department for Transport show that more than 3000 commercial vehicles are stolen in the UK every year. Only about 12% of these are ever recovered, this is because of the value in the vehicles components in addition to any load. Half of the stolen trucks are stolen from the premises of the operator.

It is worth noting that nowadays, vehicles and their loads may not only be stolen for gain but also may be used in acts of terrorism. There are many dangerous loads carried on our roads every day of the week.

What can we do to make our vehicles and their loads more secure?

Always make sure that the vehicle is left secured
- Ensure the vehicle is locked when left unattended.
- Ensure that the loading compartments are locked other than at loading or unloading times.
- Check that alarms and other security devices (if fitted) are working correctly.
- Carry a mobile phone.
- Never leave your keys in the vehicle when unattended.

The Importance of safe parking
- Plan ahead and work out where you're going to park if you are staying away overnight.
- Park in a secure area if possible.
- Don't park where there are no other vehicles around.
- Try and park in a well lit area.
- If taking a break, park where you can see your vehicle.
- Check your vehicle for signs of any irregularities.
- Ensure that the premises where vehicles are stored are secure and vehicle keys are locked away.

Be conscious of your surroundings
- Don't place yourself at risk by picking up hitch hikers.
- Be aware if something seems unusual.
- Don't discuss your job with others, particularly those you don't know.
- If stopped by police or VOSA don't leave your vehicle until you've checked their identity.
- If in doubt then telephone the police. Keep the doors locked until all is confirmed.

Committing Crime
We all know of offences such as speeding and overloading your vehicle. However, drivers can find themselves involved in far more serious criminal actions. Those travelling overseas could either intentionally or inadvertently be carrying contraband goods or illegal immigrants into the U.K. In order to avoid being on the wrong side of the law, drivers need to take action to ensure the security of their vehicles.

Accepting Bribes
Never accept any bribes, no matter how tempting. Clearly this is illegal and once you've accepted a bribe you leave yourself open to considerably more risk.

Parking in Safe Areas
Try not to park or fill up with fuel in areas that could expose you to risk (such as fuel stops near border crossings or sea ports.)

Checking the Security of the Vehicle
It's imperative that you ensure your vehicle is secured and that no one is able to tamper with the vehicle. Check all areas including the chassis and storage boxes – anywhere where goods or people could potentially be concealed (this will include the vehicle's fuel tank).

Assisting Port and Border Authorities
If you suspect that your vehicle has been tampered with in any way you should alert the relevant authorities (as soon as possible).

The Cost of Criminality
The cost of crime can be high with drivers and operators receiving heavy fines, vehicles being impounded and even confiscated. In some cases drivers have been handed out long jail sentences in the UK and abroad. Vehicles can be stolen and as stated above only 12% are ever recovered. The rule is to be ever watchful.

Questions on Criminality and its Effects

Q1. When out for a drink in your local pub, you should...
 A. Tell people about your next journey
 B. Never discuss your work
 C. Only tell them if they ask
 D. Speak quietly when discussing your work in case someone else is listening

Q2. If travelling abroad with your vehicle, you should...
 A. Fill up with fuel near the Northern French ports to take advantage of cheaper fuel
 B. Stop off for some cheaper wine before leaving mainland Europe
 C. Avoid stopping until you are inside the channel port security area
 D. Give lifts to people trying to get to the port

Q3. When stopping for a meal break you should...
 A. Always try and park where you can see your vehicle
 B. Park in an area where your vehicle cannot be seen
 C. Ask someone to keep an eye on your vehicle
 D. Never stop for rest or breaks

Q4. When leaving your vehicle, you should...
 A. Leave the keys in the ignition
 B. Hide them under the driver's seat
 C. Remove them, lock your vehicle and keep them in your pocket
 D. Leave them in the transport office

Q5. When in a foreign port and waiting to board a ferry, you should...
 A. Try and get some sleep
 B. Go and get some food and a drink
 C. Telephone home
 D. Thoroughly check over your vehicle to ensure that it has not been tampered with

Q6. On your return journey, on arrival at the ferry port, you believe that your vehicle may have been tampered with in some way. You should…
A. Do nothing
B. Alert the authorities
C. Deal with the problem yourself
D. Ask another driver for help

Q7. When driving, you are stopped by a policeman who asks you to get out of your vehicle. You should…
A. First check and confirm their identities (Warrant Cards)
B. Ask them the time
C. Telephone your boss
D. Refuse to get out under any circumstances

Q8. When parking your vehicle overnight, you should…
A. Park it where it can't be seen
B. Try and park at the side of the road
C. Park away from other vehicles
D. Park in a well lit secure area

Q9. Someone you know offers you some money to pick up a container when you are next in Spain with your coach. You should…
A. Agree because you know them
B. Agree but ask them for more money
C. Politely refuse
D. Tell them you will do it for free

Q10. You have some spare seats on your vehicle when travelling back from France. You are asked by a hotel manager to take some additional passengers back to the UK for a cash incentive. You should…
A. Take the money and tell them to get on the coach
B. Share the money with your boss when you get back
C. Telephone your transport office and ask if its OK to do so
D. Politely refuse

VEHICLE SECURITY CHECK LIST AT BORDER CROSSINGS

When checking your vehicle it is advisable to be in a safe and well lit area where there are other people around.

Drivers are advised to check their vehicles for anything that looks unusual or out of place. Smuggled items placed on or in your vehicle can be quite small. This is in addition to the problem of illegal immigrants. Use all your senses when checking. Remember it's your vehicle and you are responsible.

Drivers Name	Vehicle Registration	Vehicle Location	Date	Time

	Checklist	Satisfactory (Tick)	Comments or concerns
Inside driver's cab	Inside lockers, door pockets, cubby holes, under and behind seats, under bed/mattress (if fitted).		
Around outside of cab	Behind wind deflectors, in or behind fuel tanks, along vehicle chassis, storage boxes, behind the cab, in or behind the front grille.		
Vehicle body/loading area	First stand back from the vehicle and look for anything unusual. Then close up, check for damage to bodywork. Ensure that nothing is disturbed or damaged including seals and locks.		
Vehicle chassis	Where possible, check under vehicle along the chassis and engine compartment.		

If you have any concerns notify the authorities immediately.

THE CARRIAGE OF PASSENGERS IN THE UK AND INTERNATIONALLY

The operation of goods vehicles over 3.5 tonnes in the U.K. is strictly governed by licence. In order to carry out business an operator must have a licence. In order to obtain a licence the operator must satisfy the licensing authority (Traffic Commissioner) that they (The Operator) can comply with a number of conditions. These include the following:

- That the operator is of good repute.
- That he/she holds the appropriate qualifications to operate the business.
- That they own or lease suitable premises to carry out their business.
- That they can provide details of the vehicles used in the operation of their business.
- That they have sufficient capital resources to undertake the maintenance and repair of the vehicle or vehicles they intend to operate.
- That they have suitable premises to undertake maintenance and repair or have a contract with another service provider.

If the Traffic Commissioner is satisfied that the above conditions can be met, the operator will then be issued with one of the following licences.

- A restricted licence (Colour Orange). This limits the operator to carry their own goods but not those of a third party.
- A standard national licence (Colour Blue). This allows the operator to carry both their own goods and those of a third party for hire and reward in Great Britain.
- A standard International licence (Colour Green). This allows the operator to carry both their own goods and those of a third party for hire and reward in Great Britain and on international journeys.

The prime function of the Traffic Commissioner is to ensure that the main priority is safety. They are required by law to ensure that all the operators and vehicles licensed in their traffic area are as safe as possible.

As part of the employers' operation and the safe use of the vehicle, the driver also has responsibilities. These relate to:

- The number of hours worked and the rest and breaks taken.
- That records are kept safely (Tachographs or logbooks) and handed to the employer as required.
- The safe use of the vehicle through:
    i. How the vehicle is driven.
    ii. How it is loaded and unloaded (Passengers and luggage).
    iii. The daily safety checks.
    iv. His/her own health.
    v. The safety of other road users.
- The driver also has responsibility for his/her passengers, ensuring that they arrive safely on time.

International Journeys
When driving on international journeys drivers need to ensure that they have the appropriate documentation with them. These include:

Documents relating to the driver:
i. Their driver's licence. (Card and Paper sections).
ii. Their passport.
iii. Health Cover. (Advisable but not compulsory).
iv. A company letter giving authorisation to use the vehicle.
v. High visibility clothing for all staff.
vi. Tachograph smart cards/print rolls or discs.
vii. Fuel cards. (Advisory but not compulsory).

Documents relating to the vehicle:
i. A certificate of motor insurance for the countries visited or driven through.
ii. Registration document.
iii. A valid MOT certificate.
iv. A copy of the original community authorisation i.e. international operator's licence.
v. A check list to show that you have checked and secured your vehicle against theft or the stowage of illegal's or contraband.
vi. If towing, trailer certification (detailing ownership of the trailer).
vii. Log book, also known as a journey book or waybill.
viii. A full list of passengers.

Documents relating to the passengers:
Passengers generally have responsibility for themselves and are required to have their own documentation ready for checking where necessary. These include:
   i. Passports.
   ii. Travel insurance.
   iii. Entry visas (Non-European countries).

Drivers should also note that the list above is not complete and the operator should take advice and check the regulations of the countries that they propose to travel through en route and the final destination.

REMEMBER:
- Drink drive limits are generally lower in most European countries. And in some countries drivers are required to carry a breathalyser.
- Most motorways have toll charges.
- Warning triangles must also be carried in case of breakdown.

Questions on the Carriage of Passengers in the UK and Internationally

Q1. When driving in the UK whose responsibility is it to ensure that the vehicle is driven safely?
   A. The operator
   B. The police
   C. VOSA
   D. The driver

Q2. The colour of a restricted operator's licence as displayed in the vehicle cab is...
   A. Green
   B. Blue
   C. Orange
   D. Red

Q3. One responsibility that a driver must undertake is...
   A. To ensure that the vehicle is maintained correctly
   B. To ensure that the vehicle is insured
   C. To undertake daily safety checks
   D. To ensure that passenger have their passports

Q4. When travelling on international journeys the driver should...
   A. Ensure that his/her vehicle is secure at all times
   B. That he/she speaks the language of the country they are travelling to
   C. Ensure he/she has enough fuel for the whole journey
   D. Ensure that they have their birth certificate with them

Q5. The main purpose for the driver taking the appropriate health cover on international journeys is...
   A. To ensure that they have cover should they become ill
   B. To show the authorities that they have such cover
   C. To stop the boss going on at them
   D. To meet any legal requirements

Q6. When travelling on international journeys the driver needs to carry his/her drivers licence to...
   A. Show where they live
   B. Show proof of identity
   C. Show that they are qualified to drive that vehicle
   D. Show that they have a picture of themselves

Q7. When driving on journeys within the UK, the driver must...
   A. Ensure that they get to their destination as fast as possible
   B. Hold the appropriate driver's licence for that vehicle
   C. Carry an MOT certificate for that vehicle
   D. Carry a letter giving authorisation to use the vehicle

Q8. Drivers working within the UK have a legal requirement to ensure that they...
   A. Take the appropriate rest periods
   B. Give their employers their tachograph sheets or download their smart cards on a daily basis
   C. Ensure that all their passengers are wearing their seatbelts before they move off
   D. Carry a fuel card

Q9. When carrying passengers on international journeys, drivers must ensure that their passengers...
   A. Have appropriate medical insurance
   B. Have their driver's licences with them
   C. Have their passports
   D. Are safe whilst travelling on their vehicle

Q10. Whilst travelling on International journeys drivers are responsible for...
A. Buying their passengers drinks in the bar
B. The paperwork relating to the vehicle
C. Ensuring the passengers get the best hotels available
D. Ensuring the passenger have the correct currency for that country

UNDERSTANDING THE RISKS WHEN WORKING AND DRIVING

According to the World Health Organisation, the following number of people are killed or injured worldwide each year in road traffic accidents: -

- 1.2 Million People killed. The equivalent of a small city.
- 50 Million People Injured. The equivalent of a medium size country.

The UK now has one of the lowest death tolls per head of population in the world, with about five (5) people dying on our roads every day. That's around 1,700 per year. However if 1,700 people a year were killed in the UK due to aircraft falling out the sky there would be a public outcry with demonstrations on the streets. However, we seem to accept this human cost as inevitable and it needn't be. Much can be done to reduce the number of casualties on our roads.

Some of the latest full years data (2013) released by the Department for Transport (DfT) for the UK is shown below.

Bus and coach occupants				
	2011	2012	2013	% Change Between 2012 &2013
Fatal	7	11	10	-17%
Killed & Serious injury	332	323	342	6%
Slight injury	5845	4911	4531	-8%
All severities	6177	5234	4873	-7%

Pedestrians				
	2011	2012	2013	% Change Between 2012 &2013
Fatal	453	420	398	-5%
Killed & Serious injury	5907	5979	5396	-10%
Slight injury	20291	19239	18637	-3%
All severities	26198	25281	24033	-5%

Goods vehicle occupants				
	2011	2012	2013	% Change Between 2012 &2013
Fatal	62	62	58	-6%
Killed & Serious injury	535	561	539	-4%
Slight injury	5379	5312	5195	-2%
All severities	5941	5873	5734	-2%

Car occupants				
	2011	2012	2013	% Change Between 2012 &2013
Fatal	883	801	785	-2%
Killed & Serious injury	9225	9033	8426	-7%
Slight injury	115699	110675	101361	-8%
All severities	124924	119708	109787	-8%

Source: Department for Transport, 2014.

The cost of road traffic accidents involving injury
The cost in human terms is incalculable to those involved. The cost to the rest of us in monetary terms is estimated to be in excess of a staggering (15 billion pounds in 2013).

The cost involved not only includes the costs of the emergency services and hospitals but also the insurance costs and the loss of income and taxes from the victims.

Risk.
The biggest problem with driving is getting drivers and passengers to understand the risk they take each time they get into a vehicle.

As seen above the number of people dying in road traffic accidents is far higher than flying, yet many people perceive that there is a greater risk from flying. The internationally renowned risk expert John Adams said that understanding risk is not rocket science it's far more complicated than that. There is actual risk and perceived risk. The greatest risk comes from ourselves and the way we drive. Human behaviour is a contributory factor in over 92% of all road traffic accidents. It is precisely because of this that vehicle manufacturers are moving towards more and more automated systems such as brakes that apply themselves.

Speed
A number of studies have found that speed is the biggest factor when it comes to surviving a road traffic accident. For example one study found that in a crash when travelling at 50 mph you are 15 times more likely to die than in a crash when you are travelling at 25 mph. An Australian study found that when travelling at 60 kph (37 mph) for each additional 5 kph (3 mph) you double your risk of dying in a road traffic accident.

Casualty Reduction
During the 1990's the number of fatalities in the USA fell by 6.5 %. In the UK the number fell by 34%. Why? The USA had also benefited from improved vehicle design with seat belts, air bags and other safety features, so what made the difference in the UK? A number of factors contributed to the reduction in casualties in the UK. These included the reduction in the speed of traffic, better equipped and trained emergency services and the increase in traffic volumes.

Where do accidents happen?
Many people believe that motorways are the most dangerous roads. Again we are back to perceived risk. In fact our motorways are our safest roads when assessed against other types of road. Most fatal accidents take place on single carriageway 'A' or better 'B' class roads where the speed limit is 60 mph and there is opposing traffic travelling at the same speed. One study found that only 5% of fatal accidents take place on roads where traffic is travelling in the same direction.

When do accidents happen?
Some people think that the greatest risk is during the rush hour because there is far more traffic on the road. This is true. However these crashes tend to be low speed and low injury. One study found

that 8 out of every 1,000 crashes during normal hours resulted in a fatality.

Whereas during the rush hour 3 out of 1,000 resulted in a fatality. The reason there are fewer severe injuries is that speeds are generally lower due to heavier traffic and (usually) sober drivers at that time of day.

Who is most at risk?
Insurance companies will tell you that the premiums offered to young (particularly male) drivers are much higher due to their higher risk. A fact which most young men dispute, but none the less is true. An 18 year old male is far more likely to be involved in an accident than his mother. Drivers who have been drinking or whose driving is impaired through drugs are again far more likely to be involved in road traffic accidents. The nightmare scenario is the drunk 18 year old male driving a fast car.

Are some vehicles safer than others?
New vehicles have more safety systems fitted as standard and therefore you would think that newer cars are safer than older cars. A study in Norway which looked at 200,000 accidents found that people driving newer cars had more accidents than people driving older cars. Similar research was carried out in the USA with similar results. The researchers gave two possible answers for their findings. 1. Because new cars have more safety features, drivers felt safer so they drove faster which resulted in more accidents. 2. Drivers of new vehicles tended to drive more miles than the drivers of older vehicles and therefore had a greater chance of being involved in an accident.

What are the chances of an accident happening to me?
As we have seen, if you travel a lot on single carriageway 'A' class roads at high speed, particularly having consumed alcohol, your risk of dying in a traffic accident is quite high. That said, a famous study by American David Solomon found that if you are sober and drive too slowly, your chances of an accident are also high, as you can cause conflict with other drivers. A number of studies have also found that most road traffic accidents take place within 3 miles of the driver's home. Psychologists believe that this is due to the driver being more confident, knowing the roads and feeling safer encouraging them to take more risks.

Use of Mobile Phones and Delayed Reaction Time

Research undertaken by the Transport Research Laboratory in 2014, shows clearly how using a mobile device such as a phone when driving can endanger lives by causing a delay in the drivers' reaction time.

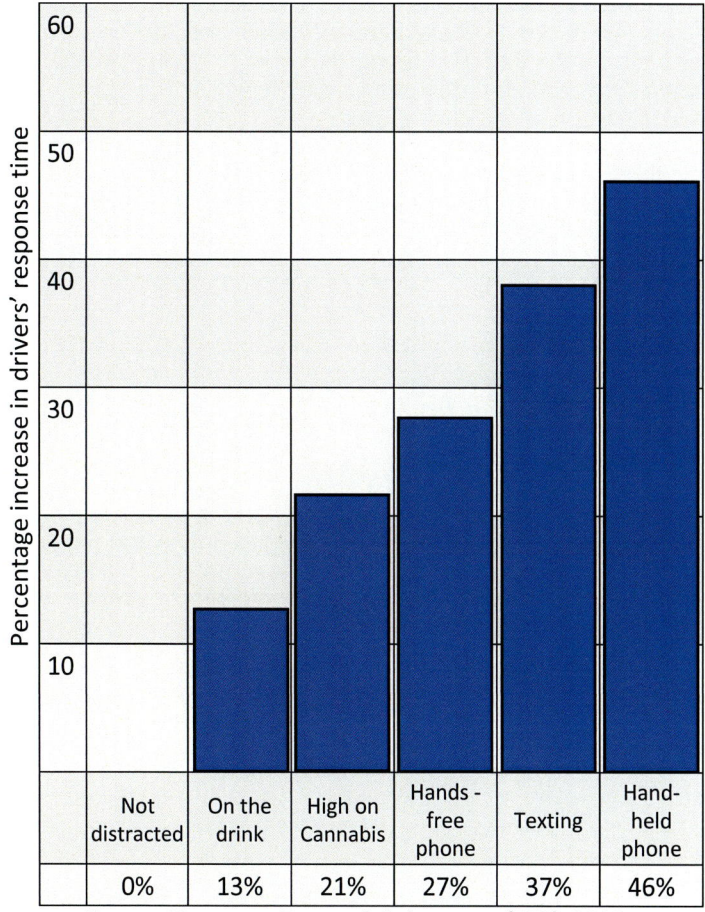

	Not distracted	On the drink	High on Cannabis	Hands-free phone	Texting	Hand-held phone
Percentage increase in drivers' response time	0%	13%	21%	27%	37%	46%

Source: Transport Research Laboratory (TRL), 2014.

Typical driver reaction time for most drivers who are not distracted or under the influence of drugs or alcohol is 1 second (shown here as 0%).

The graph shows that reaction times can increase by almost 50% when using a handheld mobile phone. So if travelling at 50 mph a driver and his vehicle will have travelled an additional 36 feet before reacting to a problem in front of them.

TRL's research found that the number of divers seen using mobiles had doubled in the 3 years between 2009 and 2012. Research undertaken by the RAC also found that almost 50% of drivers aged between 18 and 24 admitted using a mobile whilst driving.

The Department for Transport (Dft) has produced data from 2012 which shows that there were 378 road traffic incidents in which mobile phones were being used these incidents resulted in 548 casualties including 17 deaths. However, some experts believe that the figure is much higher and that in-vehicle distractions led to 9,012 incidents and 196 deaths.

In the UK using a mobile whilst driving has been an offence for some time. However, in August 2013 the penalties were increased, a fixed penalty notice is now 3 penalty points and a £100.00 fine. If a driver is convicted in a Court of Law the fine can be increased up to £1,000 and 3 penalty points.

Some road safety groups are calling for a yearlong ban if caught using a mobile whilst driving. Organizations such as RoSPA and the AA are campaigning in order to try and get the government to increase the penalties increased for using an electronic mobile device whilst driving.

Questions on Understanding Risk when Working

Q1. Approximately how many car occupants die on UK roads each year?
   A. 1,000
   B. 1,700
   C. 4,000
   D. 5,000

Q2. One of the main causes of road traffic accidents is...
   A. Driving during the rush hour
   B. Women drivers
   C. Driving too fast for the conditions
   D. Driving without using a seat belt

Q3. Young male drivers tend to...
   A. Have new faster cars
   B. Be safer drivers
   C. Have more accidents
   D. Have lower car insurance costs

Q4. Approximately how many pedestrians are injured on UK roads in 2013?
   A. 10,300
   B. 24,033
   C. 20,200
   D. 50,500

Q5. People who drive older vehicles are more likely to have an accident...
   A. True as the vehicles are less safe
   B. Only in bad weather
   C. Only men
   D. Untrue as drivers of newer vehicles tend to have more accidents

Q6. One factor in most road traffic accidents is...
   A. Poor weather conditions
   B. The poor condition of our roads
   C. Human error
   D. Poor street lighting

Q7. Where do most fatal road traffic accidents take place...
   A. On motorways
   B. On country lanes
   C. In towns and cities
   D. On single carriageway 'A' class roads

Q8. The World Health Organisation claims that the number of people killed worldwide each year is...
   A. 100,000
   B. 500,000
   C. 1 million
   D. 1.2 million

Q9. One of the safest forms of travel is...
   A. Travelling by bus
   B. Travelling by car
   C. Travelling by plane
   D. Travelling on a motorcycle

Q10. It is believed that the number of fatal accidents on UK roads has fallen due to...
   A. Better roads
   B. Faster cars
   C. Better drivers
   D. Lower traffic speeds

REPORTING A ROAD TRAFFIC INCIDENT

If you are unfortunate enough to be involved in a road traffic incident it is important that you record as much detail as possible about the event. This is not an easy task as many people may be in shock or suffer injury as a result of the incident or are simply so upset that they forget to take the other party's details.

If the incident involves injury, there are things you must do by law even if the incident was not your fault:

- Report the incident to the police at the earliest opportunity or within 24 hours.
- Give the other party or parties your details. These include:
    i. If you are the driver, your name.
    ii. Details of your vehicle.
    iii. Your contact details.
    iv. The name of your insurance company.
    v. The name and contact details of the owner of the vehicle if not you.

If the incident did not involve any obvious injury there is no requirement to report the matter to the police, unless you feel that a crime has been committed. However, it is important to get the details of the other parties and any witness to the incident.

If you have a camera (or a mobile phone with a camera), take as many pictures as possible. These need to include:

- The other vehicle or vehicles
- Damage to any property/buildings
- The road layout.

Incident Report Form

Other Vehicle Details

Reg no.		Make		Model	
Owner name		Address of registered keeper	

Details of Driver

Name		Address	
Telephone			
Mobile		E-mail	

Brief details of incident

Date		Time	
Location	
Brief details of incident	
Sketch of location and position of vehicles	

Details of witnesses

Name		Telephone	
Address		E-mail	
		Mobile	

Areas of damage (circle on vehicle)	Vehicle type	Van	LGV	PCV

Brief description of damage

Details of any other vehicles involved

Reg no.		Make		Model	
Owner name		Address of registered keeper	
Telephone			
Signature of policy holder	

ENHANCING THE IMAGE OF YOUR COMPANY

Company Image
As a driver it is important that you understand your position within the company, what your responsibilities are and how what you do can affect the company you work for. The way you conduct yourself, not only through your driving but in many other ways, plays an important part in how the company you work for is seen. Your behaviour and attitude, positive or negative is seen by the public, the customers you have contact with and other road users. It is said that you only get one chance to create a first impression.

Brand Image
The Image that companies portray to their customers, potential customers and the public at large is vital to the success of that company. Many larger companies spend millions of pounds every year on company or brand image. Just look at the cost of sign writing the vehicle you drive. Drivers are the very public face of the company and need to be aware that the public who are all potential customers are not just judging them but also the company they work for and any brand associated with that organisation. In these days of mass communication, any driving errors or bad behaviour will be seen all over our televisions and will impact on the company and other employees.

Presentation
What we look like, the way we dress and personal hygiene, all create an image not only of ourselves, but also of the company. We need to ensure that we dress appropriately for the job. Some companies issue their drivers with uniforms. A coach company may ask its drivers to wear a tie. If you are carrying scrap metal or live animals it would probably be more appropriate to wear overalls. If they get dirty try and keep a clean set in the cab. Use gloves when doing dirty jobs.

Communication
The way we speak, our tone of voice, what we say, how we say things, facial expressions, our body language and gestures are all ways of communicating positive or negative messages to customers or potential customers. Sometimes self control can be difficult. If a

customer is raising their voice try and keep calm and lower your voice. You will find that it helps matters. Shouting back won't solve anything.

Driver Behaviour
Reckless or bad behaviour of drivers or any employee does nothing to help that person, company or the customer. It can only lead to a loss of confidence in that person's ability and ultimately to their dismissal from that job and that's not in anyone's interest. The greatest asset any company can have is its staff.

What do you need to do?

Adopt safe driving practices
The way you drive can be seen by others not just the police or VOSA. Your vehicle may be sign written. If it is, you are not anonymous, unlike most car drivers. Also you may well have a "welldriven" poster on the back of your vehicle. If you are a bus or coach driver you could well have up to 60 people on board your vehicle judging your driving standards. Would you want your passengers or other road users to be frightened or put at risk because of your driving?

Develop a positive attitude to others
Not always easy, but the way we treat others is vital to our individual success in life. Think of someone you like, ask yourself why you like them. What is it about their personality? It is usually because they give of themselves; are generally happy people and ask for nothing or little in return.

Give the best customer care that you can
The customers are vital to the success of any business. Without them the business won't exist. Try and be helpful even when the customer is being difficult, not easy I know. It may be they have a point, perhaps something has gone wrong and it may not be your fault but you represent the company so try and help them if you can. Try and treat them with respect even if you think they don't deserve it.

Try and promote the company, yourself and the industry in a positive way
Being helpful and respectful to other road users, customers and the public goes a long way to achieving a positive image for all. It will also bring with it dividends.

Ensure that you load your vehicle safely
As the driver you are responsible for your vehicle safety. When loading, driving or unloading don't put yourself or others at risk by your actions or inactions. Don't take chances with your own or other people's lives, well being or goods.

Ensure that you don't take risks with the people you are carrying on your vehicle
If you are a bus or coach driver, the safety of the people on your vehicle is your number one priority and responsibility. Not just their physical safety but also any stress through mental anguish. They may be concerned because they think that you may be driving too fast or in a reckless manner. They won't come back if they have been scared, but they will tell their friends.. Make one mistake and you could find yourself on youtube within the hour.

Take pride in your job and be professional at all times
People will respect you for being the best you can. They may not tell you to your face but they will have respect. Not just the customers, but also your colleagues. You all know who the best drivers are and you know what you think of them. So why not be like them or at least try.

Questions on Enhancing your Company Image

Q1. When driving you are held up in traffic and will be late getting to your destination. Should you...
   A. Break speed limits to try and catch up
   B. Pull over and contact your office or the customer to explain the situation
   C. Do nothing
   D. Try and find a short-cut

Q2. A customer is upset with your company and makes his feelings known to you. Should you...
   A. Tell him where to get off
   B. Not bother going there again
   C. Listen to his concerns then contact your boss
   D. Listen but do nothing

Q3. You hear another employee criticising the company in front of a customer. Should you...
   A. Support him
   B. Draw him away from the customer
   C. Have an argument with him in front of the customer
   D. Do nothing

Q4. Your vehicle is dirty, do you...
   A. Forget about it
   B. Clean it yourself
   C. Wait for the next driver to do it
   D. Wait until the boss tells you to do it

Q5. You have been asked by a customer to overload your vehicle to get the job done. Should you...
   A. Agree to his request
   B. Ask your boss first
   C. Politely refuse
   D. Telephone VOSA

Q6. When driving your vehicle you see another driver driving badly. Should you...
   A. Ignore him and let him go on his way
   B. Blast him with your horn
   C. Telephone the police
   D. Make a hand gesture in his direction.

Q7. You have ripped your uniform badly whilst loading your vehicle. Should you...
   A. Ask your employer for a new one when back at base
   B. Do nothing
   C. Stop wearing a uniform
   D. Borrow one from another employee

Q8. You have been late getting your vehicle loaded (luggage) and still need to secure it. Should you...
   A. Do a quick job on securing the load
   B. Not bother securing the load
   C. Still secure the load safely before moving
   D. Secure the load and then drive fast to make up time

Q9. You are first to arrive at the scene of an accident. Should you...
    A. Stay in your vehicle and do nothing
    B. First telephone emergency services then try and help if you can and it's safe to do so
    C. Telephone your boss and tell him what's happened
    D. Telephone your other half to say you will be late

Q10. Your company has provided you with a set of uniforms. Should you...
    A. Ensure that they are all cleaned and ironed ready for work
    B. Use the same one for weeks to save washing
    C. Not use them because you don't like their colour or style
    D. Wash but not iron and hope the creases come out

ECONOMICS AND THE TRANSPORT INDUSTRY

As we have all seen in recent years with the banking crisis, economics plays an important part of all our everyday lives. Many companies became insolvent and people lost their jobs through no fault of their own.

The matter may well be out of our hands at a national or international level. However, it helps to try and understand what is going on closer to home and what we can do to ensure that the company we work for makes a profit which, at the end of the day, allows us to keep our jobs and hopefully have pay rises. We all understand that if the company we work for is making a loss, then the company and our jobs are at risk.

Unique Selling Point
Everything we buy, be it an item or a service, has a 'Unique Selling Point'. That means it has something that sets it apart from the rest. Generally, this is what prompts us to buy that item or service. Examples of why something has a unique selling point are set out below:
- **Price.** We buy items from one supplier because they're cheaper than elsewhere.
- **Quality.** We buy because we believe that the quality is better than elsewhere.
- **Good service.** We believe that the service is better than elsewhere.
- **Availability.** It may not be available elsewhere or it will take longer to arrive.
- **Convenience.** We can't be bothered to go elsewhere.
- **Relationships.** We know the person we are dealing with and like them.

Drivers are in the front line of any service. We may not be able to set the price but we do influence our customers in many other ways. It is important that we understand that we need to ensure that the customer is happy and we are giving the best possible service we can. Developing a positive attitude and going that extra yard has a direct affect on the business we work for and will more than likely, keep us in a job.

Company Costs

All companies have costs. These can be placed in two categories, fixed cost and variable costs.

Fixed Costs: These are the costs which are set and are usually in place for one year. They include things like the costs of insurance, road tax, finance on vehicles and wages of full time employees (not overtime.) It is important here to understand that if we don't turn up for work our employers still have these costs. A bus or truck sitting on the yard is still costing the owners money. There may also be issues with the loss of business if you are not there..

Variable Costs: These costs can change during the year. They include things like the cost of fuel, tyres and the maintenance of vehicles.

Breakeven Chart: Set out below is a simple break-even chart which shows how much business a company needs to do in order to make a profit.

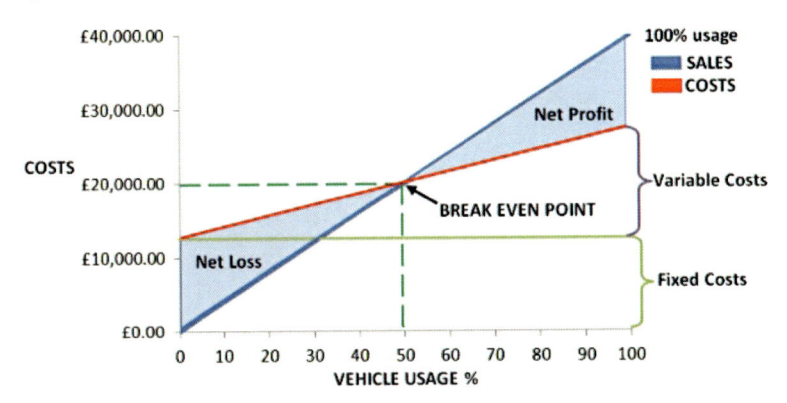

Fixed Costs: We can see at the bottom of the chart that our fixed costs are set. As stated above these include:

- Insurances (vehicle insurance, public liability insurance and employer's liability insurance).
- Wages and national insurance of all company employees.
- Rent and loans on buildings and vehicles.
- Council Tax on buildings.
- The tax on vehicles.

These fixed costs will stay the same throughout the year unless the

the employer takes on more staff, bigger premises or more vehicles. If the employer expands the business, then the fixed costs will rise.

Variable Costs: This is shown as the cost line. We can see the more work that is carried out the higher this line moves. Variable costs include:

- Fuel costs. The more miles we do the more fuel we buy.
- Maintenance costs. Again the more miles we do the more maintenance is required.
- Overtime and temporary staff. We may have so much extra work that we have to take on temporary or additional staff.

Break-even Point: It can be seen from the chart that the business does not break-even until we reach 50% capacity. It is only after this point that we move into profit. So if any business loses work it will have a dramatic impact on that business. It should also be noted that if that business is not charging enough it may not break-even until well beyond the 50% mark.

So how much does it cost?
An example of the costs of running one mid-size commercial vehicle per year is set out here. However, if your employer has 100 vehicles then these costs will be 100 times higher. Also if you have newer or bigger vehicles then the costs are going to be even more.

Section 1

Drivers wages including holiday and NI	£22,000	Fixed Cost
Vehicle insurance	£3,000	Fixed Cost
Vehicle tax	£3,000	Fixed Cost
Fuel	£25,000	Variable Cost
Maintenance including tyres	£7,000	Variable Cost
Vehicle finance	£8,000	Fixed Cost
Depreciation	£7,000	Variable Cost
Total	£75,000 per annum	

It should be noted that the company has other additional costs. These include the following:

Section 2
- Wages of office staff.
- Employer's liability insurance.
- Public liability insurance.
- Council tax on buildings.
- Heating and lighting of buildings.
- Cost of the buildings (mortgage or rent).

Calculating Costs
Calculating the cost of hiring out one of your vehicles is an important task. Get it wrong and charge too much and you won't get the business. Charge too little and you could be doing the work at a loss. So how much should you charge?

Looking at the annual running costs of a mid size vehicle shown above, we can work out the running cost per mile. Then if we know where the customer wants us to go, we can calculate the total mileage (not forgetting the return journey).

We can see that our annual fuel costs are £25,000. So let's keep it simple and say that the cost of each litre of fuel before VAT is £1.00. If we are getting 3 miles per litre that means that our vehicles total mileage is 75,000 miles per year.

To then get the running cost per mile we should divide the total cost of running the vehicle (£75,000) by the total annual mileage (75,000 miles):

£75,000 / 75,000 miles = cost per mile of £1.00 per mile.

Now we have that figure we need to add in our other costs shown in Section 2.

Questions Economics and the Transport Industry

Q1. Vehicle insurance can be considered to be...
    A. A fixed cost
    B. A variable cost
    C. An unimportant cost
    D. Profit for the company

Q2. If a vehicle is parked on the yard and not run it is...
  A. Not costing the company any money
  B. Saving the company money as it is not using any fuel
  C. Costing the company money
  D. There to make the yard look busy

Q3. As a driver, I have no part to play in keeping the customer happy.
  A. This is true
  B. I am the face of the company and have significant influence on how that business is seen by the customer
  C. That's not my job, I'm just the driver
  D. If the customer's not happy that's his problem

Q4. Any point above the break even point on the chart is considered to be...
  A. A profit
  B. A way of avoiding tax
  C. A loss
  D. A way of saving money

Q5. You are told that your company is not making money during the current economic climate, you believe that...
  A. It should only concern the bosses
  B. It should concern everyone in the company
  C. It should only concern those who joined the company last
  D. The company should cut back on things like insurance.

Q6. If as a driver I speed and get to my destination faster it will...
  A. Make more profit for the company
  B. Have no effect
  C. Put lives at risk and use more fuel
  D. Improve my standing in the company

Q7. Maintenance costs are considered to be...
  A. A fixed cost
  B. A way of reducing costs
  C. Nothing to do with me as a driver
  D. A variable cost

Q8. The cost of the fuel for the vehicle I drive is determined by how fast I go.
   A. True
   B. It has no effect
   C. False
   D. Only if I get caught speeding

Q9. If a company charges too much for its services...
   A. It would have no effect on how profitable it is
   B. It could lose customers
   C. It could attract more customers
   D. It could make far too much money

Q10. If a company charges too little for its services...
   A. It could lose customers
   B. It will make a loss
   C. It could make a bigger profit
   D. It will have to pay more tax

Q11. Which of the following is a unique selling point?
   A. Poor service
   B. Selling damaged goods cheaply
   C. Good customer care
   D. Late delivery of items

Q12. As a driver it is not part of my job to keep the customer happy.
   A. This true, it's the boss's problem
   B. Only if I get paid extra money
   C. I'm paid to drive not keep customers happy
   D. Not true, customer care is my responsibility

APPENDICES

LICENCE CATEGORIES FOR DRIVING PCVS

You may drive PCVs if you have the following categories on your licence and comply with the notes.

Licence Category	Description of Vehicle	Minimum Age	Notes
D1	Minibuses. Vehicles with between 9 and 16 passengers' seats and with a trailer of up to 750 kg.	21	**NB** You may drive aged 17 if a member of the armed forces. You may drive aged 18 if one of the following applies: 1. You are learning to drive this category of vehicle or driver CPC initial qualification. Or if you are undergoing a national vocational training course to obtain a CPC initial qualification. 2. You have passed your driving test and driver CPC Initial qualification and are driving on a regular service where the route does not exceed 50 km. Or if you are not engaged in the carriage of passengers. 3. You have had your driving licence before 10 September 2008. However you must take your driver CPC before September 2013. 4. Driving under a bus operator's licence, minibus permit or community bus permit. **NB** If you have passed your test for category B or B automatic before 1st January 1997 (not for hire and reward).
D1 + E	Minibuses with Trailers. This is a combination of vehicles as in category D1 where the trailer has a maximum mass of 750kg. The maximum mass of the vehicle and trailer must not exceed 12,000 kg. Also the maximum authorised mass of the trailer does not exceed the unladen weight of the towing vehicle.	21	

Licence Category	Description of Vehicle	Minimum Age	Notes
D	Buses. Any bus with more than 8 passenger seats and with a trailer up to 750 kg.	21	**NB** You may drive aged 17 if a member of the armed forces. You may drive aged 18 if one of the following applies. 1. You are learning to drive this category of vehicle or driver CPC initial qualification. Or if you are undergoing a national vocational training course to obtain a CPC initial qualification. 2. You have passed your driving test and driver CPC Initial qualification and are driving on a regular service where the route does not exceed 50 km. Or if you are not engaged in the carriage of passengers. 3. You have had your driving licence before 10 September 2008. However you must take your driver CPC before September 2013. 4. Driving under a bus operator's licence, minibus permit or community bus permit. **NB** You need a category D licence to drive an articulated bus (e.g. bendy bus).
D + E	Buses with Trailers. Any bus with more than 8 passenger seats and with a trailer of more than 750 kg	21	

For further information contact the DVLA or read leaflet INS57P.

SPEED LIMITS

Type of Road Type of vehicle	Built-up areas* Speed in mph (km/h)	Single carriage ways Speed in mph (km/h)	Dual carriage ways Speed in mph (km/h)	Motorways Speed in mph (km/h)
Cars & Motorcycles (including car-derived vans up to 2 tonnes maximum laden weight)	30 (48)	60 (96)	70 (112)	70 (112)
Cars towing caravans or trailers (including car-derived vans and motorcycles)	30 (48)	50 (80)	60 (96)	60 (96)
Buses, coaches and minibuses (not exceeding 12m in overall length)	30 (48)	50 (80)	60 (96)	70 (112)
Goods vehicles (not exceeding 7.5 tonnes maximum laden weight)	30 (48)	50 (80)	60 (96)	70 (112)
Goods vehicles (exceeding 7.5 tonnes maximum laden weight)	30 (48)	40 (64)	50 (80)	60 (96)

GUIDE FOR DRIVERS
THE 'COUNTY TICKET' AND SCHOOL BUS PASSES

This year, we are issuing a greatly increased number of bus tickets for use on the school bus network. You will be seeing two different types of ticket.

1 The County Ticket
This is issued to students attending Colleges of Further Education and Sixth Forms within schools. A few County Tickets have also been issued to pupils who are younger, but you won't see these on school buses. **Only** those County Tickets that have a school transport route number in the bottom right hand corner, e.g. C999, are to be allowed on to school buses.

2 School Bus Pass
These are issued to all pupils travelling to all middle, secondary and sixth form schools. These bus passes also have the route number shown prominently in the bottom right hand corner.

The two types of ticket look like these examples but the ticket colour will change annually.

County Ticket   School Bus Pass

Please note the route number or numbers – these should match the route that you are driving.

You must check tickets every day for every journey. If you do not, the County Council could fine your company quite heavily.

Please do not refuse school pupils entry to your bus just because they are not carrying their School Bus Pass – they could come to serious harm if left on their own. If a pupil turns up without a ticket, take his or her name and report it to your manager. Only refuse entry if your manager or this office instructs you to do so.

If you have any doubts, please contact your manager.

ANSWERS TO DRIVER CPC QUESTIONS FOR PCV DRIVERS LICENCES D, D + E, D1 AND D1 + E

Working Time Directive.
(1.C) (2.C) (3.D) (4.C) (5.B) (6.D) (7.C) (8.D) (9.D) (10.C)

Drivers Hours.
(1.D) (2.D) (3.B) (4.C) (5.C) (6.B) (7.B) (8.C) (9.B) (10.A) (11.A) (12.D) (13.A) (14.C) (15.A)

Tachographs.
(1.A) (2.A) (3.B) (4.C) (5.A) (6.B) (7.B) (8.D) (9.C) (10.C) (11.A) (12.D) (13.B) (14.D) (15.B)

Understanding your vehicle and its characteristics.
(1.B) (2.C) (3.A) (4.C) (5.C) (6.D) (7.A) (8.C) (9.C) (10.B)

More efficient use of fuel.
(1.B) (2.A) (3.C) (4.C) (5.A) (6.C) (7.B (8.C) (9.C) (10.B)

Vehicle safety and walk-around checks.
(1.D) (2.B) (3.A) (4.A) (5.B) (6.C) (7.C) (8.C) (9.A) (10.A)

Safety of wheels and tyres.
(1.C) (2.B) (3.A) (4.D) (5.D) (6.D) (7.B) (8.D) (9.B) (10.A)

Dimensions and weights of vehicles.
(1.C) (2.D) (3.B) (4.B) (5.B) (6.D) (7.C) (8.C) (9.C) (10.D)

The operation of the transmission system.
(1.C) (2.C) (3.C) (4.A) (5.B) (6.D) (7.A) (8.A) (9.D) (10.B)

The operation of the braking system.
(1.C) (2.A) (3.C) (4.B) (5.D) (6.C) (7.D) (8.A) (9.D) (10.D)

Health and wellbeing when driving.
(1.C) (2.B) (3.C) (4.C) (5.D) (6.A) (7.C) (8.A) (9.B) (10.D)

First aid.
(1.B) (2.C) (3.C) (4.B) (5.A) (6.D) (7.B) (8.C) (9.C) (10.B)

Manual Handling.
(1.B) (2.C) (3.B) (4.A) (5.C) (6.C) (7.D) (8.A) (9.B) (10.B)

Passengers Safety & comfort.
(1.A) (2.C) (3.C) (4.B) (5.C) (6.D) (7.C) (8.B) (9.C) (10.B)

The forces affecting your vehicle and its passengers.
(1.D) (2.B) (3.C) (4.B) (5.B) (6.B) (7.C) (8.C) (9.A) (10.D)

Seatbelt Regulations.
(1.B) (2.C) (3.C) (4.D) (5.B) (6.D) (7.C) (8.A) (9.C) (10.D)

The carriage of people with disabilities or special needs.
(1.C) (2.D) (3.B) (4.C) (5.A) (6.C) (7.D) (8.A) (9.B) (10.C)

How to assess and deal with emergency situations.
(1.B) (2.C) (3.B) (4.D) (5.C) (6.B) (7.B) (8.C) (9.D) (10.C)

Evacuation Procedures for Buses & Minibuses
(1.C) (2.C) (3.B) (4.A) (5.D) (6.C) (7.C) (8.D) (9.A) (10.D)

Managing your passengers and conflicts.
(1.C) (2.B) (3.C) (4.C) (5.D) (6.C) (7.A) (8.C) (9.B) (10.B)

Dealing with criminality and its effects.
(1.B) (2.C) (3.A) (4.C) (5.D) (6.B) (7.A) (8.D) (9.C) (10.D)

Understanding the regulations governing the carriage of passengers in the UK and internationally.
(1.D) (2.C) (3.C) (4.A) (5.A) (6.C) (7.B) (8.A) (9.D) (10.B)

Understanding the risks when working and driving.
(1.B) (2.C) (3.C) (4.B) (5.D) (6.C) (7.D) (8.D) (9.C) (10.D)

Company Image.
(1.B) (2.C) (3.B) (4.B) (5.C) (6.A) (7.A) (8.C) (9.B) (10.A)

Economics and the transport industry.
(1.A) (2.C) (3.B) (4.A) (5.B) (6.C) (7.D) (8.A) (9.B) (10.B) (11.C) (12.D)

INDEX

A
Acceleration.. 83
Axel
    Weights... 43

B
Brake retarders... 52, 53
    Electronic.. 52
    Exhaust brake function..................................... 52
Braking.. 83
Braking systems... 51, 52
    ABS and electronic.. 51
    Air brakes... 52
    Hydraulic.. 51
    Parking brake... 52
Brand image.. 131
Breaks.. 4, 6-8, 11, 12

C
Child restraints... 87
Comfort... 82
    Passenger.. 76, 81
Company image.. 131, 133
    Customer care.. 132
    Driver behaviour.. 132
Conflict
    Handling..................................... 57, 103, 104, 108
Cornering.. 83
County Ticket... 146
CPC.. 1-3
    Non-UK citizens... 2
Criminality... 111-113
    Bribes.. 112
    Committing.. 112
    Cost... 112
    Security check list.. 115
    Victims.. 111
Customer care.. 132

D
Delayed Reaction Time..................................... 125, 126
Disabled/Special needs passengers................... 90-100
Documents
    Driver.. 117
    Passengers... 118

Vehicle.. 117
DRABC technique.. 63-64
Drink and drug driving... 57, 59
    Limits and penalties... 58
drivercpc... 3
Drivers' hours.. 9
    Domestic... 9, 13
    EU... 9, 10
    Working day... 9, 10
    Working week... 10
Drivers' hours... 17

E

Economics... 136, 137, 139
    Breakeven chart.. 137
    Breakeven point... 138
    Fixed costs... 137
    Unique selling point.. 136
    Variable costs... 137, 138
Emergency situations.. 94-95
    Break down... 95
    Fire.. 95, 96
    Road traffic accident... 94
    Seriously ill passenger.. 95
Evacuation procedure... 98-101
Eyesight, *See* Health

F

Fatalities.. 121
Fatigue.. 59, 60
First Aid... 63, 65-68
    Broken bones... 68
    CPR... 65
    DRABC.. 63
    Heart attack.. 67
    Kit... 63
    Recovery position.. 66
    Severe bleeding... 66
    Shock.. 66, 67
Fitness to drive... 77
Forces... 81, 83, 84
    Centrifugal.. 83
    Friction... 83
    Gravity... 83
Fuel

Consumption.. 30
Efficient use.. 30
Saving... 30

G

Gearbox.. 48
Fully automatic... 48
Modern.. 48

H

Health... 55-60
Eyesight... 55
Healthy eating.. 60
Medicals.. 55
Health assessments... 6
Height of vehicle.. 27, 44
Holiday... 4-6
Horizontal Amending Directive... 4

I

Incident report form... 130
Injuries... 121
Cost... 122
International and UK
Carriage of passengers... 116

J

Journey planning.. 77

L

Legal Offence... 126
Length of vehicle... 27-44
Licence
Categories... 143, 144
Loading.. 78, 79, 82, 133

M

Manual handling.. 71-73
Mobile Phones... 125, 126
Multi-manning.. 9, 14, 23

N

Night Work... 7
Night work... 5

O

Operator licenses... 116

P

Port and border authorities... 112
Security check list... 115

R

Rest period.. 6, 8
Rest periods... 12, 13
Risks.. 121, 123, 124, 133
Road rage.. 57, 108
Road traffic incident... 129, 130
    Incident report form.. 130
    Reporting.. 129
Road Transport Directive.. 4, 6, 8

S

Safety.. 33, 78, 79
    Control.. 77
    Evacuation procedure.. 98
    Passenger... 76, 77
    Vehicle.. 124, 126
School Bus Passes.. 146
School runs... 104-106
Seat Belts.. 87
    Exemption certificate.. 87
    Regulation.. 86, 87
Seat belts.. 86
    Penalties.. 87
Securing vehicles... 111, 112
    Parking.. 111, 112
Smoking.. 59
Speed limits... 145
Stress.. 55-57

T

Tachographs... 17, 18
    Analogue.. 17, 19, 20
    Digital.. 17, 21, 22
    Inspection.. 18
    Manual Entry... 20
    Smart cards... 21, 23
Traffic commissioner... 116
Transmission systems.. 47, 48
Travelling... 83

U

Unforseen events.. 11

V

Vehicle defect report... 38

W

Walk around check list...................................... 36, 37

Walk around checks.. 33, 77
Weight of vehicle... 27, 43
Wheels and tyres.. 39, 40
    Checks... 39, 40
    Safety of.. 39, 40
    Wheel nut indicators... 40
    Wheel nuts... 40
Width of vehicle.. 27, 44
Wind... 84
Working time
    Recording.. 23
Working Time Directive... 4, 5
Working Week.. 4-7
Working Time... 4
    Recording.. 6